The Enteric Microbiota

Colloquium Series on Integrated Systems Physiology: From Molecule to Function to Disease

Editors

D. Neil Granger, *Louisiana State University Health Sciences Center-Shreveport*
Joey P. Granger, *University of Mississippi Medical Center*

Physiology is a scientific discipline devoted to understanding the functions of the body. It addresses function at multiple levels, including molecular, cellular, organ, and system. An appreciation of the processes that occur at each level is necessary to understand function in health and the dysfunction associated with disease. Homeostasis and integration are fundamental principles of physiology that account for the relative constancy of organ processes and bodily function even in the face of substantial environmental changes. This constancy results from integrative, cooperative interactions of chemical and electrical signaling processes within and between cells, organs, and systems. This eBook series on the broad field of physiology covers the major organ systems from an integrative perspective that addresses the molecular and cellular processes that contribute to homeostasis. Material on pathophysiology is also included throughout the eBooks. The state-of the-art treatises were produced by leading experts in the field of physiology. Each eBook includes stand-alone information and is intended to be of value to students, scientists, and clinicians in the biomedical sciences. Since physiological concepts are an ever-changing work-in-progress, each contributor will have the opportunity to make periodic updates of the covered material.

Published titles

(for future titles please see the Web site, www.morganclaypool.com/page/lifesci)

The Enteric Microbiota
Francisco Guarner
www.morganclaypool.com

ISBN: 9781615041985 paperback

ISBN: 9781615041992 ebook

DOI: 10.4199/C00047ED1V01Y201110ISP029

A Publication in the

COLLOQUIUM SERIES ON INTEGRATED SYSTEMS PHYSIOLOGY: FROM MOLECULE TO FUNCTION TO DISEASE

Lecture #29

Series Editors: D. Neil Granger, LSU Health Sciences Center, and Joey P. Granger, University of Mississippi Medical Center

Series ISSN

ISSN 2154-560X print

ISSN 2154-5626 electronic

The Enteric Microbiota

Francisco Guarner
Digestive System Research Unit, University Hospital Vall d'Hebron

COLLOQUIUM SERIES ON INTEGRATED SYSTEMS PHYSIOLOGY: FROM MOLECULE TO FUNCTION TO DISEASE #29

MORGAN & CLAYPOOL LIFE SCIENCES

ABSTRACT

The human gut is the natural habitat for a diverse and dynamic microbial ecosystem having an important impact on health and disease. Bacteria have lived in and on animal hosts since multicellular life evolved about 1 billion years ago. Hosts provide habitat and nutrition to the microbial communities and derive many benefits from their guests that contribute with metabolic (recovery of energy and nutrients), defensive (barrier effect against invaders) and trophic (immune regulation, neuro-endocrine development) functions. Several disease states or disorders have been associated with changes in the composition or function of the enteric microbiota, including inflammatory bowel diseases, obesity and the metabolic syndrome. Probiotics and prebiotics can be used to improve symbiosis between enteric microbiota and host, or correct states of dysbiosis.

KEYWORDS

microbiology, evolution, human development, human physiology, colon, probiotics, prebiotics, immune system, intestinal inflammation, obesity, metabolic syndrome, inulin, oligofructose, galacto-oligosaccharides.

Contents

CHAPTER 1

Microbial Communities

Human beings are associated with a large and diverse population of microorganisms that live on body surfaces and in cavities connected with the external environment. Whereas researchers have paid more attention to pathogenic microbes, i.e., microorganisms able to invade and induce infectious diseases in humans and animals, chronic microbial colonization that inflict no evident harm on their hosts only attracted minor scientific attention during the past century (Moran et al., 2008). However, vertebrate and invertebrate animals are in permanent association with such microbial communities maternally inherited at birth or acquired from the environment during the first stages of life (Moran et al., 2008). Associations that benefit the host as well as the microbe are grouped under the term "symbiosis" and the microbial partners called "symbionts." The prevalence of symbiosis has long been recognized on the basis of observations from microscopy, but most aspects of symbiont origins and functions have remained unexplored before the age of molecular techniques because of the difficulties to cultivate and isolate a large majority of these microbial species. The present work focuses on our current knowledge about microbial symbionts in humans. Development of novel gene sequencing technologies as well as availability of powerful bioinformatic analysis tools have allowed a dramatic proliferation of research studies over the past few years.

Evidence indicates that the skin, mouth, vagina, upper respiratory tract and gastrointestinal tract of humans are inhabited by site-specific microbial communities with specialized structure and functions (Dethlefsen, 2007; Costello, 2009). "Microbiota" is a collective term for the microbial communities in a particular ecological niche, and this expression is preferred rather than "flora" or "microflora," which perpetuate an outdated classification of bacteria as plants (Dethlefsen et al., 2006). Thus, the term "enteric microbiota" refers to the ecosystem of microorganisms which have adapted to live on the intestinal mucosal surface or within the gut lumen (Guarner & Malagelada, 2003; Ley et al., 2006a; Dethelfesen, 2007). The human enteric microbiota is highly diverse in composition and includes microbes belonging to all three domains of life on Earth, i.e., Bacteria, Archaea and Eucarya (Figure 1), as well as their phages and viruses.

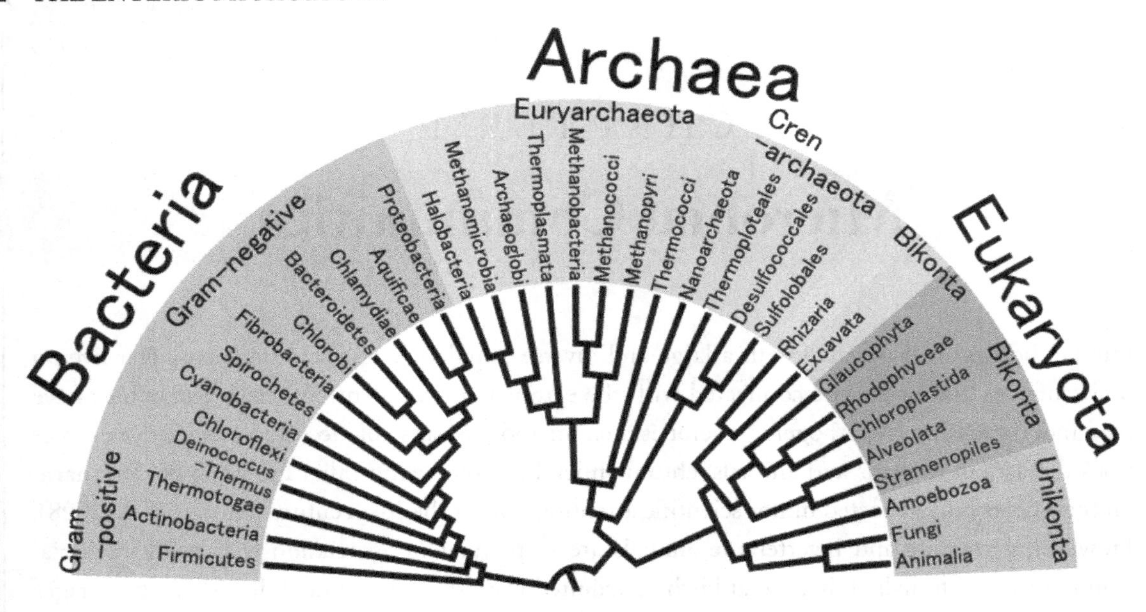

FIGURE 1: The Tree of Life is based on comparisons of nucleotide sequences in ribosomal RNA genes. The ribosomal RNA genes are highly conserved genes and are universally present in living organisms. The diagram compiles the results of phylogenetic analysis by comparisons of many rRNA sequences. Source: Wikimedia Commons http://en.wikipedia.org/wiki/File:Phylogenetic_Tree_of_Life.png.

1.1 LIFE ON EARTH

Bacteria have been on Earth for 3.5 billion years, appearing approximately 1 billion years after the Earth's crust was formed (Schopf, 2006). Fossils and associated geochemical markers of biologic activity indicate that microbial organisms inhabited the oceans in Archean times (2.5 to 3.7 billion years ago). Photomicrographs in Figure 2 show microfossils of structurally preserved filamentous microbes that were discovered in northwest Western Australia (Schopf, 1993). Gas chromatography-mass spectrometry studies of biologic lipids extracted from bitumen shales of the same Australian region identified substances that are characteristic biomarkers of cyanobacteria (Brocks et al., 1999). These studies suggested that early microbial communities synthesized hydro-carbonated compounds and were capable both of photosynthetic oxygen production and respiratory oxygen consumption (Schopf, 2006). The presence early in Earth history of morphologically cyanobacterium-like fossils has been widely assumed to be the origin of free oxygen gas in the atmosphere, suggesting that both oxygenic photosynthesis and aerobic respiration of eukaryotic cells are processes derived from microbial biochemistry (Schopf, 2006). Geochemical biomarkers suggest that the domain Eucarya first appeared 2.7 billion years ago (Brocks et al., 1999), though fossil records of eukaryotes are dated 0.5 to 1 billion years after that time.

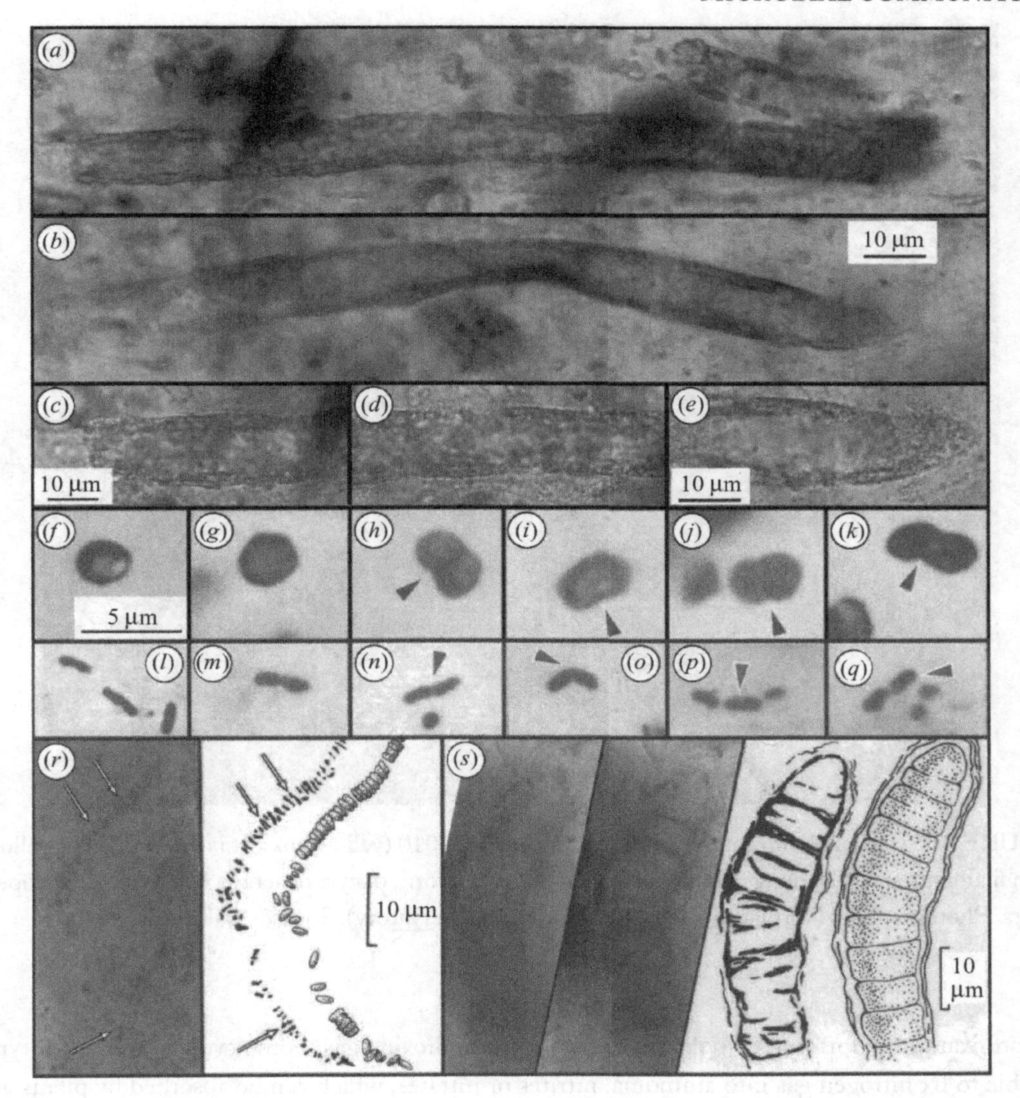

FIGURE 2: Microfossils of cyanobacterium-like microorganisms photographed in petrographic thin sections of Archean rocks from South Africa and Western Australia. Source: from Figure 3 in: Schopf JW. Fossil evidence of Archaean life. *Philos Trans R Soc Lond B Biol Sci.* 2006;361: 869–885. Reproduced with permission of The Royal Society.

Cyanobacteria are still vastly abundant in modern days, and can be found as planktonic cells in oceans and fresh water (Figure 3). They also occur in damp soil or on moistened rocks. They do not require organic nutrients and can grow on entirely inorganic materials (Stewart & Falconer, 2008). Cyanobacteria obtain their energy through photosynthesis and convert solar energy into biomass-stored chemical energy. Like plants, the cyanobacteria release oxygen gas and contribute to

FIGURE 3: Bacterial bloom south of Fiji on October 18, 2010 (yellow box). It is likely that the yellow-green filaments are miles-long colonies of *Trichodesmium*, a form of cyanobacteria often found in tropical waters. Photograph by Norman Kuring (NASA Earth Observatory). Source: Wikipedia.

carbon fixation by forming carbohydrates from carbon dioxide gas. Some cyanobacteria cell types are able to fix nitrogen gas into ammonia, nitrites or nitrates, which can be absorbed by plants and converted to protein and nucleic acids (nitrogen gas is not bioavailable to plants). Cyanobacteria fulfill vital ecological functions and have a massive impact across the planet, being important contributors to global carbon and nitrogen cycles (Stewart & Falconer, 2008).

Microbial communities are ubiquitous and truly essential for maintaining life conditions on Earth. As summarized in a report from a colloquium convened by the American Academy of Microbiology, microbial communities can be found in every corner of the globe, from the permafrost soils of the Arctic Circle to termite guts in sub-Saharan Africa, and on every scale, from microscopic biofilms to massive marine planktonic communities (Buckley, 2002). Because of their enormous global size, microbial communities have a massive impact across the globe. Their diverse contributions affect every aspect of life, from human infections, to the treatment of chemical contamination, to the cycling of the most critical elements for maintaining life. Evolution, disease, corrosion, degradation, bioremediation and global cycling are a few of the many thousands of ways in which the impact of microbial communities is felt (Buckley, 2002).

1.2 PROKARYOTIC CELLS

Bacteria and Archea are prokaryotes, i.e., unicellular organisms that do not have a cell nucleus, mitochondria or any other membrane-bound organelles, and are usually much smaller in size than eukaryotic cells. The genome of prokaryotic cells is held in the cytoplasm without a nuclear envelope and consists of a single loop of stable chromosomal DNA, plus other satellite DNA structures called plasmids that are mobile genetic elements and provide a mechanism for horizontal gene transfer within a population (Figure 4). In contrast, DNA in eukaryotes is found on tightly bound and organized chromosomes, not suitable for horizontal gene transfer.

Genome size and number of coding genes are much smaller in prokaryotes than in eukaryotes (Hou & Lin, 2009) (Figure 5). Genome size is a gross estimate of biological resources linked to a given species and correlates with a range of features at the cell and organism levels, including cell size, body size, organ complexity and extinction risk. Thus, single microbial species may not have enough genetic resources by their own for adequate fitness and survival. Single species are likely to have obligate dependencies on other species, including other microbes or host animals or plants

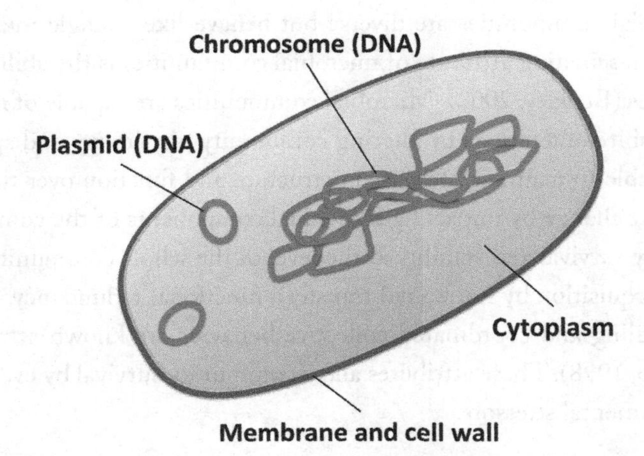

FIGURE 4: Prokaryotic cells do not have a nucleus, mitochondria or any other membrane-bound organelles. The genome is held in the cytoplasm without a nuclear envelope and consists of a single loop of stable chromosomal DNA, plus other satellite DNA structures called plasmids. Source: Drawing by the author.

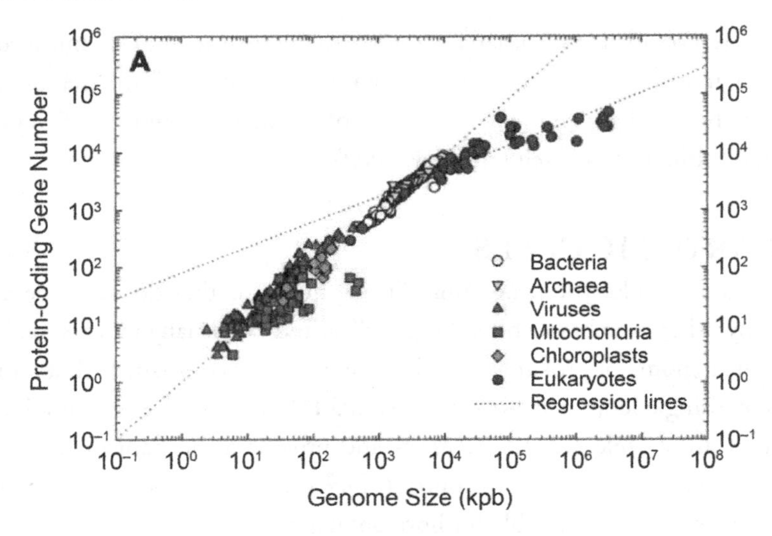

FIGURE 5: Genome size and number of protein-coding genes of eukaryotes, bacteria, archea, viruses and organellas are shown on logarithmic scales. Genome size is much smaller in prokaryotes than in eukaryotes. Source: from figure 2A in: Hou Y, Lin S (2009). Distinct gene number-genome size relationships for eukaryotes and non-eukaryotes: gene content estimation for dinoflagellate genomes. Source: PLoS ONE 4 (9): e6978. (open access)

(Moran et al., 2008). Therefore, multispecies communities with complex nutritional and social interdependencies are the natural lifestyle for survival for most prokaryotic micro-organisms.

Natural microbial communities are diverse but behave like a single multicellular organism (Shapiro, 1998). One fascinating attribute of microbial communities is the ability for adaptation to environmental changes (Buckley, 2002). Microbial communities are capable of recovering from and adapting to radical habitat alterations by altering community physiology and species composition. In this way, they are able to maintain stability in structure and function over time. Environmental alterations bring about change by impacting individual components of the community, but the results are manifested by survival and stability at the level of the whole community. Genetic diversity and plasticity (gene acquisition by horizontal transfer), functional redundancy, metabolic cooperation, cell-to-cell signaling and coordinated collective behavior are known attributes of microbial communities (Shapiro, 1998). These attributes allow community survival by evolving, adapting and responding to environmental stressors.

1.3 INTESTINAL MICROBIAL ECOLOGY

A beneficial role on host health by microbial communities of symbionts was already suspected in early times of modern microbiology. Louis Pasteur is best remembered for his remarkable breakthroughs in demonstrating the germ theory of disease, as well as for developing strategies, such as vaccines and pasteurization, to prevent and combat infections. With his awareness of the causative role of some specific microbes in producing disease, the prominent French scientist presumed that other microbes would be essential for life. Pasteur suggested that animals would not be able to survive when totally deprived of "common" microorganisms (Pasteur, 1885).

During the twentieth century, researchers developed appropriate facilities and technologies to breed experimental animals under absolute "germ-free" conditions in order to study the impact of microbial colonization on host physiology (Wostmann, 1981). These studies demonstrated that, contrary to Pasteur's presumption, animal life is possible in the absence of microbial colonization. However, a major challenge to achieving survival in the germ-free state was to develop adequate diets to meet the extraordinary nutritional requirements in the absence of microbial colonization (Wostmann, 1981). In addition, germ-free animals did not develop normally in terms of body anatomy and physiology (Figure 6). These intriguing findings attracted growing attention from scientists during the last decades of the past century. Microbial colonization of animals may not be essential for life, but the microbiota is critical for normal growth and development.

Although all epithelial surfaces of mammalians are colonized by microorganisms, the gastrointestinal tract has the largest microbial burden. In humans, the gastrointestinal tract houses

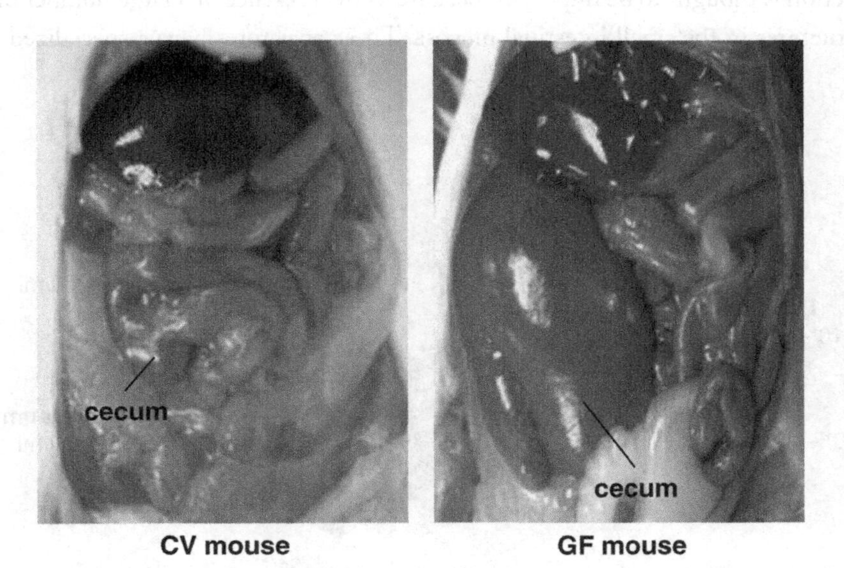

FIGURE 6: The abdominal cavity of germ-free (GF) mice is enlarged as compared with mice bearing a conventional (CV) microbiota. The distended cecum is a prominent characteristic of GF mice, as evident in the left photograph. The small intestine of GF mice is usually shorter and the liver smaller than in CV mice. On the other hand, body fat content is higher in CV than in GF mice. Source: pictures from the author's lab.

over 10^{14} microbial cells with over 1000 diverse microbial species, most of them belonging to the domain Bacteria (Qin et al., 2010). Microbial communities in the gut include native species that colonize the intestine permanently, and a variable set of living microorganisms that transit temporarily through the gastrointestinal tract. On the other hand, the mucosa of the gastrointestinal tract constitutes a major interface with the external environment and is the body's principal site for interaction with the microbial world. The gastrointestinal mucosa exhibits a very large surface (considering the villus-crypt structure in an unfolded disposition, a flat extension of up to 4,000 square foot is estimated) and contains adapted structures and functions for bi-directional communication with microorganisms, including a number of preformed receptors, microbial recognition mechanisms, host–microbe cross-talk pathways and microbe-specific adaptive responses (Cummings et al., 2004; MacDonald et al., 2011).

The stomach and duodenum harbor very low numbers of microorganisms, typically less than 10^3 bacteria cells per gram of contents, mainly lactobacilli and streptococci (Figure 7). Acid, bile and pancreatic secretions suppress most ingested microbes, and phasic propulsive motor activity impedes stable colonization of the lumen. The numbers of bacteria progressively increase along the jejunum and ileum, from approximately 10^4 cells in the jejunum to 10^7 cells per gram of contents in the distal ileum. In the upper gut, transit is rapid and bacterial density is low, but the impact on immune function is thought to be important because of the presence of a large number of organized lymphoid structures in the small intestinal mucosa. These structures have a specialized epithelium

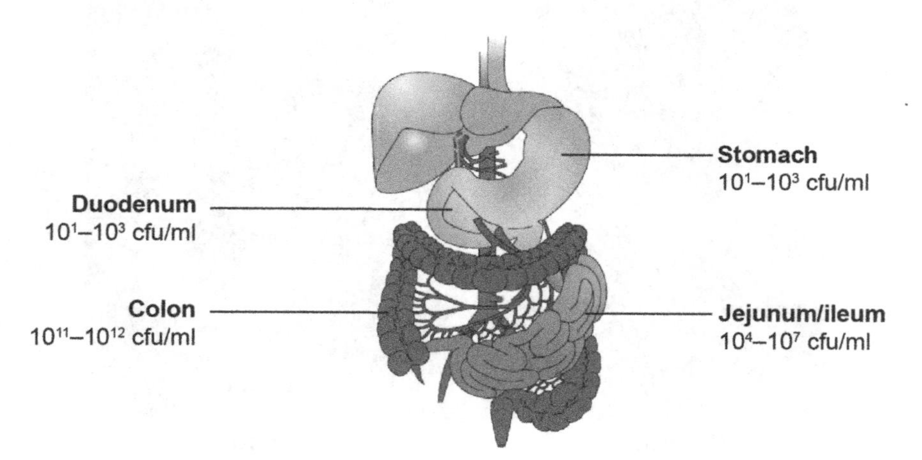

FIGURE 7: The gastrointestinal tract houses over 10^{14} microbial cells with over 1,000 diverse microbial species, most of them belonging to the domain Bacteria. The large intestine is the most densely populated habitat due to the slow transit time and the availability of fermentable substrates. Source: Figure 1a in O'hara AM, Shanahan F (2006). The gut flora as a forgotten organ. *EMBO Rep* 7: 688–93. Reproduced with permission of Nature Publishing Group.

for uptake and sampling of antigens and contain lymphoid germinal centers for induction of adaptive immune responses (Yamanaka et al., 2003).

In the colon, however, transit time is slow and microorganisms have the opportunity to proliferate by fermenting available substrates derived from either the diet or endogenous secretions. The large intestine is heavily populated by anaerobes with up to 10^{12} cells per gram of luminal contents. By far, the colon harbors the largest population of human microbial symbionts, which contribute to 60% of solid colonic contents (O'Hara & Shanahan, 2006). Several hundred grams of bacteria living within the colonic lumen certainly affect host homoeostasis.

• • • •

CHAPTER 2

Host–Microbe Interactions in the Gut

Until recent years, information on enteric microbiota in medical and scientific books was rather scarce. Certain bacteria of the human gut microbiota are associated with toxin formation and pathogenicity when they become dominant. Some other resident species are potential pathogens when the integrity of the mucosal barrier is functionally breached. Knowledge on overt or latent pathogens advanced markedly, due to their ability to translocate to the blood stream or other body sites, the infections or clinical complications they can cause, or the virulence factors they carry.

However, the normal interaction between gut bacteria and their host is a symbiotic relationship, defined as mutually beneficial for both partners. The host provides a nutrient-rich habitat, and intestinal bacteria confer important benefits on the host's health (Hooper et al., 2002). Surprisingly, knowledge on enteric bacteria with proven benefits for human health is very rudimentary. Several beneficial features of gut bacteria are widely recognized, including production of short chain fatty acids, vitamin synthesis, secretion of defensins or bacteriocins and inhibition of pathogens through a multiplicity of mechanisms (Guarner & Malagelada, 2003; O'Hara & Shanahan, 2006). However, there is currently little consensus regarding definition or characterization of potentially healthy bacteria in the human gut. Thus, our current concepts on host–microbe symbiosis in the gut are mainly supported by observations using germ-free animal models.

2.1 PRIMARY FUNCTIONS OF THE ENTERIC MICROBIOTA

Comparison of animals bred under germ-free conditions with their conventionally raised counterparts (conventional microbiota) has revealed a series of anatomic characteristics and physiological functions that are associated with the presence of the microbiota (Wostmann, 1981; Falk et al., 1998). As mentioned previously, germ-free animals have extraordinary nutritional requirements in order to sustain body weight and are highly susceptible to infections. Organ weights (heart, lung and liver), cardiac output, intestinal wall thickness, gastrointestinal motility, serum gamma-globulin levels, lymph nodes, among other characteristics, are all reduced or atrophic in germ-free animals. Recent data have also shown that germ-free mice display greater locomotor activity and reduced anxiety when compared with mice with a normal gut microbiota (Diaz Heijtz et al., 2011). Reconstitution of germ-free animals with a microbiota restores most of these deficiencies, suggesting that gut bacteria provide important and specific tasks to the host's homeostasis (Figure 8). Evidence obtained through such animal models suggests that the main functions of the microbiota are ascribed into three categories, i.e., metabolic, protective and trophic functions (Guarner & Malagelada, 2003; O'Hara & Shanahan, 2006).

Metabolic functions consist of the fermentation of nondigestible dietary substrates and endogenous mucus. Gene diversity among the microbial community provides a variety of enzymes and biochemical pathways that are distinct from the host's own constitutive resources. Fermentation of carbohydrates is a major source of energy in the colon for bacterial growth and produces short chain fatty acids that can be absorbed by the host. These biochemical conversions result in salvage of di-

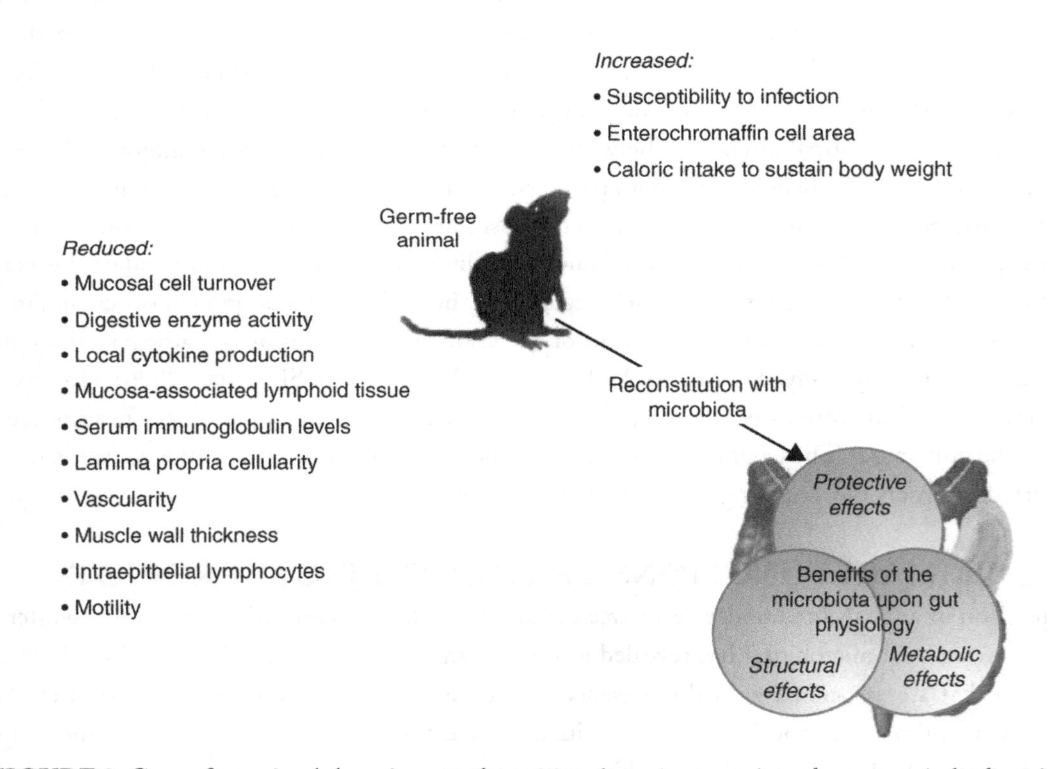

FIGURE 8: Germ-free animals have increased nutritional requirements in order to sustain body weight, are highly susceptible to infections and show structural and functional deficiencies. Reconstitution of germ-free animals with a microbiota restores most of these deficiencies, suggesting that gut bacteria provide important and specific tasks to the host's homeostasis. Source: Figure 1 in O'Hara AM, Shanahan F. Gut microbiota: mining for therapeutic potential. *Clin Gastroenterol Hepatol* 2007 Mar;5(3): 274–84. Reproduced with permission of Elsevier.

etary energy and favor the absorption of ions (Ca, Mg, Fe) in the cecum. Colonic microorganisms also play a role in vitamin synthesis (Conly et al., 1994).

Protective functions of gut microbiota include the barrier effect that prevents invasion by pathogens. Resident bacteria represent a resistance factor to colonization by exogenous microbes or opportunistic bacteria that are present in the gut, but their growth is restricted. The equilibrium between species of resident bacteria provides stability in the microbial population, but antibiotics can disrupt the balance (for instance, overgrowth of toxigenic *Clostridium difficile*).

Trophic functions include the control of epithelial cell proliferation and differentiation, modulation of certain neuro-endocrine pathways, and the homeostatic regulation of the immune system. Epithelial cell differentiation is influenced by the interaction with resident microorganisms as shown by the expression of a variety of genes in germ-free animals mono-associated with specific bacteria strains (Hooper et al., 2001), and in humans fed with probiotic lactobacilli (Van Baarlem et al., 2011). The microbiota suppresses intestinal epithelial cell expression of a circulating lipoprotein–lipase inhibitor, fasting-induced adipose factor (Fiaf), thereby promoting the storage of triglycerides in adipocytes (Backhed et al., 2005).

The ability of the gut microbiota to communicate with the brain and thus influence behavior is emerging as an exciting concept (Cryan & O'Mahony, 2011). Recent reports suggest that colonization by the enteric microbiota impacts mammalian brain development and subsequent adult behavior. In mice, the presence or absence of conventional enteric microbiota influences behavior and is accompanied by neurochemical changes in the brain (Neufeld et al., 2011). Germ-free mice have increased locomotor activity and reduced anxiety, and this behavioral phenotype is associated with altered expression of critical genes in brain regions implicated in motor control and anxiety-like behavior (Diaz-Heijtz et al., 2011). When germ-free mice are reconstituted with a microbiota early in life, they display similar brain characteristics as conventional mice. Thus, the enteric microbiota can affect normal brain development.

Gut microbes also play an essential role in the development of a healthy immune system (Round & Mazmanian, 2009). Animals bred in a germ-free environment show low densities of lymphoid cells in the gut mucosa and low levels of serum immunoglobulins. Exposure to commensal microbes rapidly expands the number of mucosal lymphocytes and increases the size of germinal centers in lymphoid follicles (Yamanaka et al., 2003; Bouskra et al., 2008). Immunoglobulin producing cells appear in the lamina propria, and there is a significant increase in serum immunoglobulin quantities. Most interestingly, recent findings suggest that some commensals play a major role in the induction of regulatory T cells in gut lymphoid follicles (Atarashi et al., 2011). Control pathways mediated by regulatory T cells are essential homeostatic mechanisms by which the host can tolerate the massive burden of innocuous antigens within the gut or on other body surfaces without resulting in inflammation (Guarner et al., 2006; Round & Mazmanian, 2009; MacDonald et al., 2011).

2.2 THE MICROBIOTA AS A METABOLIC ORGAN

The enteric microbiota has a collective metabolic activity equal to a virtual organ within the gastrointestinal lumen (O'Hara & Shanahan, 2006; Backhed et al., 2005). Microbial communities of symbionts are composed of different cell lineages and provide the host with a huge diversity of genes and functions. In this way, symbionts can enhance host ability to acquire nutrients from the environment or provide the pathways for synthesis of needed organic compounds or for catabolism of molecules available in the environment (Moran, 2007). For animals, the genes encoding enzymes for biosynthesis of many required organic compounds were lost early in evolution. Thus, animals are unable to make many of the compounds required for cellular function and are limited in their abilities to use different energy sources. Bacterial or fungal symbionts have evolutionary adapted to provide the required organic compounds (essential amino acids and vitamins) and the ability to obtain energy from different sources (Moran, 2007). The guts of ruminants and termites are well-studied examples of host–microbe metabolic partnership. Symbiont bacteria carry out the task of breaking down complex polysaccharides of ingested plants and provide nutrients and energy for both microbiota and host (Backhed et al., 2005; Flint et al., 2008). Moreover, the amino acid supply of ruminants eating poorly digestible low-protein diets largely depends on the microbial activities in their forestomachs (Figure 9). Interesting studies in monogastric animals, including humans, have shown that microbial synthesis of essential amino acids in the intestinal tract is used to support growth and protein homeostasis (Metges et al., 2006).

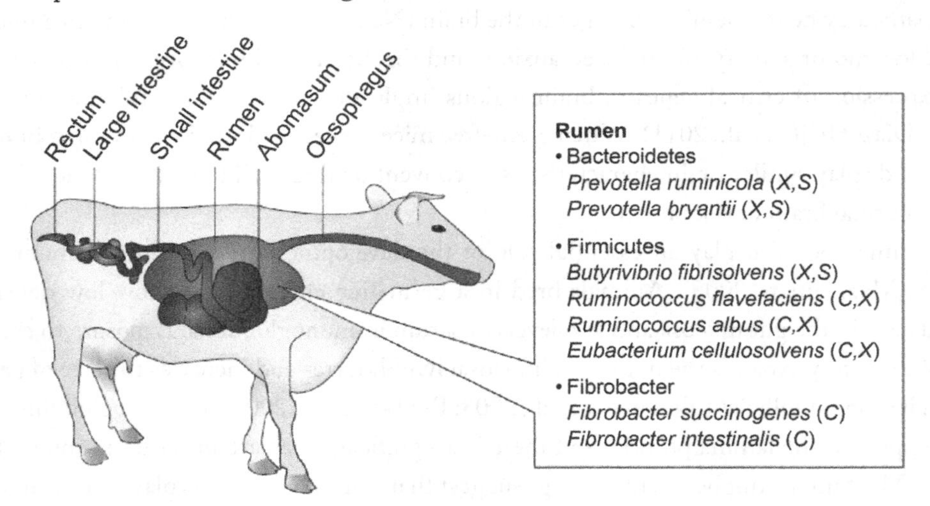

FIGURE 9: The amino acid supply of ruminants eating poorly digestible low-protein diets largely depends on the microbial activities in their forestomachs. Source: Figure 1 in Flint HJ, Bayer EA, Rincon MT, Lamed R, White BA (2008). Polysaccharide utilization by gut bacteria: potential for new insights from genomic analysis. *Nat Rev Microbiol* 6: 121–31. Reproduced with permission of Nature Publishing Group.

In the human being, the distal intestine represents an anaerobic bioreactor programmed with an enormous population of microbes (Backhed et al., 2005). Due to the slow transit time of colonic contents, resident microorganisms have ample opportunity to degrade available substrates, which consist of non-digestible dietary residue and endogenous secretions. Colonic microbial communities provide genetic and metabolic attributes to harvest otherwise inaccessible nutrients.

The vast majority of the bacteria in the colon are strict anaerobes and thus derive energy from fermentation. Substrates that reach the colon include non-digestible carbohydrates, dietary fibers, non-digested dietary proteins, other proteins of endogenous origin (enzymes, secretions, desquamated epithelial cells) and mucins, the glycoprotein constituents of the mucus which lines the walls of the gastrointestinal tract (Table 1).

Under normal physiological circumstances, the two main fermentative substrates in the colon are non-digestible carbohydrates and proteins. Lipids are only present at influential levels in

TABLE 1: Fermentable substrates that reach the human colon.

SUBSTRATE	COMPONENT	AMOUNT (g/day)
Carbohydrates	Resistant starch	5–35
	Non-digestible polysaccharides	10–25
	Oligosaccharides (e.g., fructo- or gluco-oligosaccharides, inulin)	2–8
	Monosaccharides (e.g., sugars, sugar alcohols)	2–5
	Mucins	3–5
	Synthetic carbohydrates (e.g., lactulose, polydextrose, modified cellulose)	Variable
Proteins	Dietary origin	1–12
	Endogenous origin (e.g., pancreatic enzymes and other secretions)	4–8
	Desquamated epithelial cells	30–50
Other	Non-protein nitrogen (e.g., urea, nitrate)	Around 0.5
	Organic acids, lipids, bacterial recycling	Unknown

Adapted from Egert et al. (2006).

patients with severe pancreatic insufficiency. The carbohydrate fraction that reaches the colon is variable and depends largely on the composition of the ingested food, whereas cellular desquamation is the principal source of proteins and is less variable (Egert et al., 2006). The overall amount of non-digestible carbohydrate available for colonic fermentation in subjects on a Western diet may vary from 20 to 80 g/day. Endogenous carbohydrates, chiefly from mucins and chondroitin sulphate, only contribute about 3–5 g/day of fermentable substrate. On the other hand, up to 50 g of protein are fermented in the human colon daily (Egert et al., 2006; Roberfroid, 2005).

Carbohydrates are fermented in the colon to short chain fatty acids, mainly, acetate, propionate and butyrate (Cummings, 1981), and a number of other metabolites, such as lactate, pyruvate, ethanol, succinate as well as the gases H_2, CO_2, CH_4 and H_2S (Levitt et al., 1995). Short chain fatty acids acidify the luminal pH which suppresses the growth of pathogens. They also influence intestinal motility (Cherbut, 2003) and contribute towards energy requirements of the host (Cummings, 1981). Acetate is metabolized in human muscle, kidney, heart and brain. Butyrate is largely

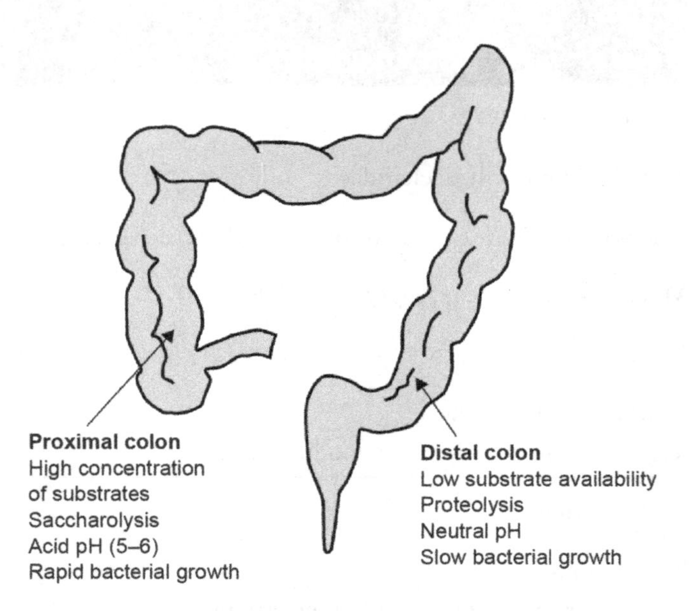

FIGURE 10: The human proximal colon is a saccharolytic environment. Fermentation of undigested carbohydrates is intense with high production of short-chain fatty acids, and rapid bacterial growth. By contrast, carbohydrate availability decreases in the distal colon and putrefactive processes of proteins and amino acids are the main energy source for bacteria. Source: Figure 1 in Guarner F, and Malagelada Jr (2003). Gut flora in health and disease. *Lancet* 361: 512–9. Reproduced with permission of Elsevier.

metabolized by the colonic epithelium where it serves as the major energy substrate as well as a regulator of cell growth and differentiation (Cummings, 1981; Williams et al., 2003).

Proteins reaching and/or produced in the colon are fermented to branched chain fatty acids such as isobutyrate, isovalerate and a range of nitrogenous and sulphur-containing compounds. Unlike carbohydrate fermentation products which are recognized as beneficial to health, some of the end products of amino acids metabolism may be toxic to the host (e.g. ammonia, amines and phenolic compounds) (Macfarlane et al., 1992).

The human proximal colon is a saccharolytic environment with the majority of carbohydrate entering the colon being fermented in this region (Figure 10). In the distal colon, carbohydrate availability decreases, and proteins and amino acids become increasingly important energy sources for bacteria (Macfarlane et al., 1992). Consequently, excessive fermentation of proteins in the distal colon has been linked with disease states such as colon cancer and chronic ulcerative colitis, which generally affect the distal region of the large intestine. Thus, it is recognized as favorable to shift the gut fermentation towards saccharolytic activity by increasing the proportion on non-digestible carbohydrates in the diet (Roberfroid et al., 2010).

2.3 THE MUCOSAL IMMUNE SYSTEM

The intestinal immune system is the largest and most complex part of the immune system. At least 80% of the antibody production of the adult human body takes place locally in the intestinal mucosa (Brandtzaeg, 2009). The vast gastrointestinal surface is continuously exposed to both potential pathogens, food antigens and indigenous commensal microorganisms travelling through the gut lumen. Indeed, the intestinal mucosa is the main interface with the external environment and is adapted with specialized structures and functions for immune recognition of microbes and foreign substances (Brandtzaeg, 2009; MacDonald et al., 2011). The mucosal immune system has evolved to provide both optimal defense against pathogens and tolerance toward dietary antigens and commensal non-pathogenic microbes (Brandtzaeg, 2010). A clear discrimination between pathogens and harmless antigens is critical for health since immuno-inflammatory reactions against foreign structures can damage the host's own tissues (Guarner et al., 2006). These notions are very relevant for a proper understanding of the interactive co-existence of the immune system and the enteric microbiota.

Studies in germ-free animals have clearly documented the key role of the microbiota for an optimal structural and functional development of the immune system (Yamanaka et al., 2003; Round and Mazmanian, 2009; Gaboriau-Routhiau et al., 2009). For instance, germ-free mice are immuno-deficient and highly susceptible to pathogen or opportunistic infections. In addition, they fail to develop normal adaptation to dietary antigens like ovo-albumin, and oral tolerance mechanisms are depressed or abrogated. These abnormalities can be corrected by reconstitution of a conventional microbiota, but this procedure is only effective in neonates and not in older mice (Guarner & Malagelada, 2003). Massive interaction between gut microbial communities and the mucosal immune compartments early in life seem to be critical for a proper instruction of the immune system. Later in life, multiple and diverse interactions between microbes, epithelium and gut lymphoid tissues are constantly reshaping local and systemic immunity.

Intestinal epithelial cells are in close contact with luminal contents and play a crucial role in signaling and mediating host innate and adaptive mucosal immune responses (Kagnoff, 2006). Activation of innate host defense mechanisms is based on the rapid recognition of conserved molecular patterns in microbes by pre-formed receptors (membrane-bound toll-like receptors and NOD-family receptors in the cytosol). In response to invading bacteria, "alarm" signals generated by these cellular receptors converge to transcription factors (NF-kappaB and others), which start the transcription of genes responsible for the synthesis of pro-inflammatory proteins. Hence, epithelial cells secrete chemoattractants for neutrophils and pro-inflammatory cytokines and express inducible enzymes for the production of inflammatory mediators (nitric oxide, prostaglandins, leukotrienes, etc.). Another important strategy for defense is minimizing contact between luminal microorganisms and the epithelial cell surface. This is accomplished through the production of mucus and antimicrobial proteins such as defensins and cathelicidins (Kagnoff, 2006).

On the other hand, epithelial cells are also able to recognize microbe-associated molecular patterns from non-pathogenic bacteria and may elicit different cytokine responses that are transmitted to underlying immunocompetent cells (Haller et al., 2000). Interestingly, responses to non-pathogenic bacteria involve regulatory cytokines, such as TGFbeta or IL-10, and appear to be related with the induction regulatory pathways of the immune system. For instance, some Lactobacillus strains can downregulate the spontaneous release of TNFalpha by inflamed tissue, and also the inflammatory response induced by *E. coli* (Borruel et al., 2002). These effects on cytokine release are associated with changes in the expression of activation markers by lamina propria T lymphocytes and with induction of apoptosis of activated lymphocytes, which is a major homeostatic mechanism in the gut mucosa. Thus, signals generated at the mucosal surface can promote changes in the phenotype of lamina propria lymphocytes. Intestinal epithelial cells can also express MHC class II and non-classical MHC class I molecules, suggesting that they may function as antigen-presenting cells. Taken together, epithelial cells produce the essential signals for the onset of mucosal innate responses and recruitment of appropriate cell populations for induction of memory pathways of acquired immunity.

Acquired immune responses develop in specialized lymphoid tissues. Intestinal immune cells are located in three compartments: organized gut-associated lymphoid tissue (GALT), the lamina propria, and the surface epithelium. GALT comprises Peyer's patches, the appendix and numerous isolated lymphoid follicles (Brandtzaeg, 2009). These structures represent inductive sites for intestinal immune responses, while the lamina propria and epithelial compartment principally constitute effector sites (Figure 11). Peyer's patches consist of at least five aggregated lymphoid follicles, and occur mainly in the ileum and less frequently in the jejunum. Human Peyer's patches begin to form during gestation but no germinal centers appear until shortly after birth, reflecting dependency on stimulation from the environment. Numbers of Peyer's patches increase from approximately 50 at

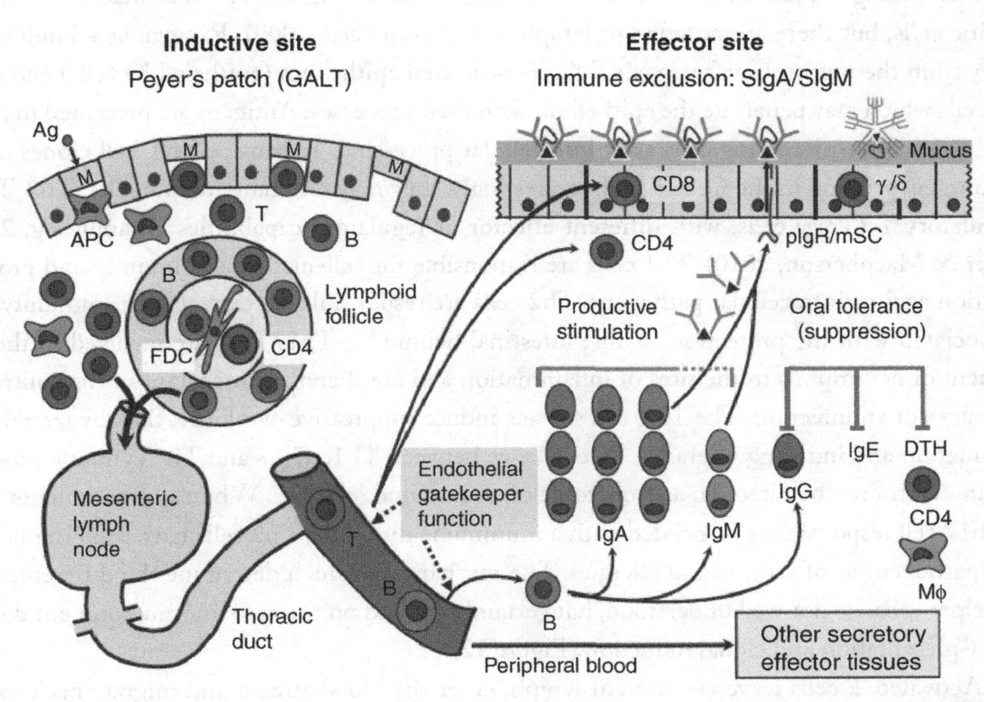

FIGURE 11: Gut-associated lymphoid tissue (GALT) comprises Peyer's patches and isolated lymphoid follicles which are inductive sites for mucosal T and B cells. Exogenous antigens (Ag) are actively transported through M-cells of the follicle-associated epithelium to reach professional antigen-presenting cells (APC), including macrophages and follicular dendritic cells (FDC). After being primed, naïve T and B cells become effector cells and migrate from GALT to mesenteric lymph nodes via efferent lymph and then via the thoracic duct to peripheral blood for subsequent extravasation at mucosal effector sites. The mucosal lamina propria (effector site) is illustrated with its various immune cells, including B cells (B), Ig-producing plasma cells, and CD4+ T cells. The distribution of intraepithelial lymphocytes (mainly CD8+ T cells) is also schematically depicted. Plasma cells mainly produce secretory IgA (SIgA) and secretory IgM (SIgM) which are delivered to the gut lumen via membrane secretory component (mSC). Source: Figure 5 in Brandtzaeg P. Mucosal immunity: induction, dissemination, and effector functions. *Scand J Immunol* 2009 Dec;70(6): 505–15. Reproduced with permission of Wiley-Blackwell.

the beginning of the last trimester to 100 at birth and 250 in young adults. The human intestinal mucosa harbors at least 30,000 isolated lymphoid follicles, increasing in density distally. Thus, they are 10 times more frequent in the ileum than in the jejunum, but the great majority of them are found in the colonic mucosa. Interestingly, the organogenesis of murine isolated lymphoid follicles was found to commence after birth, in contrast to the Peyer's patches (Bouskra et al., 2008).

The Peyer's patches and isolated lymphoid follicles resemble lymph nodes with B-cell follicles, intervening T-cell zones and a variety of antigen-presenting cells, such as macrophages and dendritic cells, but there are no afferent lymphatics (Brandtzaeg, 2009). Exogenous stimuli come directly from the mucosal surfaces via a follicle-associated epithelium (epithelial M cells) and dendritic cells which may penetrate the epithelium with their processes. Antigens are presented to naïve T cells by antigen-presenting cells after intracellular processing. Expansion of T cell clones occur after antigen-priming in the GALT and, interestingly, they may differentiate into Th1, Th2, Th17 or regulatory T (Treg) cells, with different effector or regulatory capabilities (Brandtzaeg, 2010; Hooper & Macpherson, 2010). Th1 cells are responsible for cell-mediated immunity and provide protection against intracellular pathogens. Th2 cells are responsible for extracellular immunity and are associated with the protection against intestinal helminths. Th17 cells are involved in the recruitment of neutrophils to the sites of inflammation and are therefore important in the control of early stages of an infection. The Treg cell subsets induce suppressive cytokines, thereby restraining inflammation and inducing tolerance. The balance between Th1, Th17 and Th2 cytokine production can determine the direction and outcome of an immune response. Whereas predominant Th1 and Th17 cell responses are associated with autoimmune disorders, Th2 cells have been implicated in the pathogenesis of asthma and allergies. The mechanisms which determine the differentiation of T helper cells are not well understood, but certainly depend on the cytokine environment during antigen-presentation and clonal induction (Figure 12).

Activated T cells leave via efferent lymph, enter the blood stream and migrate back to the lamina propria where they constitute a pool of activated effector cells. Primed B cells migrate via draining lymphatics to mesenteric lymph nodes where they are further stimulated; they may then reach peripheral blood and become seeded by preferential homing mechanisms into distant mucosal effector sites, particularly the intestinal lamina propria where they finally develop to plasma cells. Mucosal antigen-specific B cells differentiate to predominantly IgA-secreting plasma cells (Brandtzaeg, 2009). In the normal human gastrointestinal tract, T lymphocytes and plasma cells producing IgA (derived from antigen-primed B cells) constitute up to two thirds of the cells in the lamina propria (MacDonald et al., 2011).

Effector T cells (Th1, Th2 or Th17) that react to microbial or other luminal antigens are controlled by Treg cells. Defects in Treg cells would lead to mucosal inflammation after challenge with the antigen. The most well-studied Treg cells include T cells that express the transcription factor fork

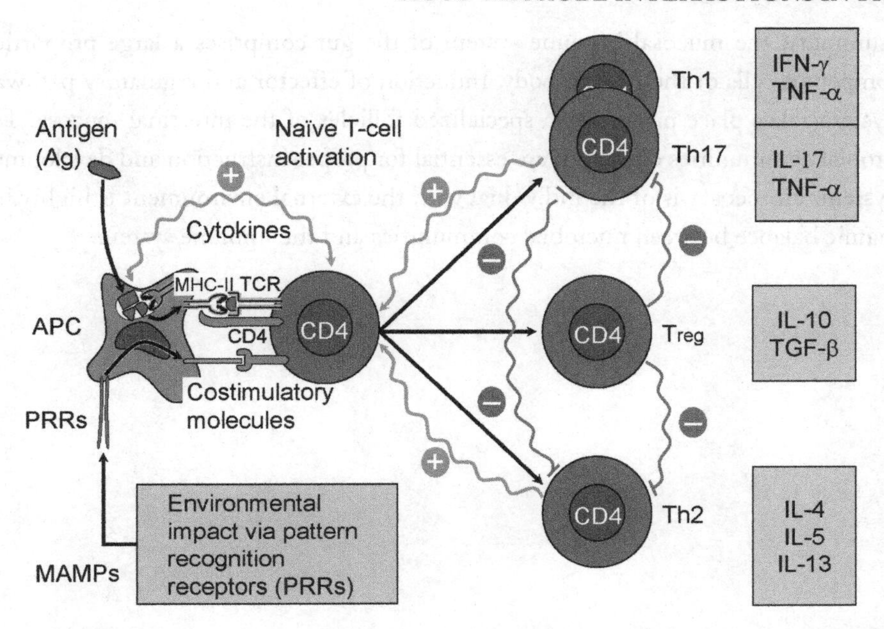

FIGURE 12: The specialized lymphoid follicles of the gut mucosa are the major sites for induction and regulation of immune responses. Gut microbes stimulate clonal expansion of lymphocytes, which may differentiate into Th1, Th2, Th17 or Treg cells, with different effector or regulatory capabilities. Experimental evidence suggests that innate recognition of microbe associated molecular patterns (MAMPs) by epithelial cells and/or antigen-presenting cells (APC) plays a decisive role for the induction of either effector or regulatory pathways. Co-stimulatory signals (cytokines) released by these cell types dictate polarization of the adaptive immune response. Source: modified from Figure 3 in: Guarner F, Bourdet-Sicard R, Brandtzaeg P, Gill HS, McGuirk P, van Eden W, Versalovic J, Weinstock JV, Rook GA (2006). Mechanisms of Disease: the hygiene hypothesis revisited. *Nat Clin Pract Gastroenterol Hepatol* 3: 275–284.

head box p3 (FoxP3). Two classes FoxP3-expressing Treg cells have been identified (MacDonald et al., 2011). The first so-called naturally occurring or constitutive Treg cells were recognized as a subpopulation of T helper cells that develop in the thymus during the first days of postnatal life and express high levels of the IL-2 receptor (CD25). Naturally occurring Treg cells account for 1% to 2% of peripheral T helper cells and are believed to maintain tolerance toward self-antigens. In contrast, inducible Treg cells develop from naïve T helper cells primed by antigen-presenting cells in the presence of regulatory cytokines (TGFbeta, IL-10). The intestinal environment appears to be an important site for generation of inducible Treg cells. After antigen-specific activation, inducible Treg cells suppress several populations of effector T cells in an antigen-independent manner. Inducible Treg cells might maintain tolerance toward dietary- and microbial-derived antigens.

In summary, the mucosal immune system of the gut comprises a large proportion of the immunocompetent cells of the human body. Induction of effector and regulatory pathways of the immune system takes place primarily in specialized follicles of the intestinal mucosa. The indigenous microbial communities of the gut are essential for proper instruction and development of the immune system. Homeostasis of the individual with the external environment is highly influenced by the dynamic balance between microbial communities and the immune system.

• • • •

CHAPTER 3

Composition of the Human Enteric Microbiota

Our current knowledge about the microbial composition of the human enteric microbiota in health and disease is still very limited. Conventional bacteriological analysis of fecal specimens requires meticulous techniques for the cultivation of bacteria on a large variety of growth media and an array of biochemical tools for taxonomic identification of the isolates. Detailed microbiologic analysis of fecal samples by conventional culture methods is a monumental task; it was estimated that bacteriologic characterization of a single specimen would take up to 1 year to complete (Simon & Gorbach, 1984). Such studies demonstrated that strict anaerobic bacteria outnumber aerobes by a factor of 100 to 10,000 (Simon & Gorbach, 1984). The predominant isolates were *Bacteroides, Bifidobacterium* and *Eubacterium*. Other common isolates were anaerobic Gram-positive cocci (*Clostridium, Peptococcus, Peptostreptococcus, Enterococus* and *Ruminococcus*) and the Enterobacteriaceae (*Escherichia, Enterobacter, Klebsiella, Proteus*, etc.).

Several interesting concepts were drawn from such culture-dependent studies. For instance, each individual harbors several hundreds of species belonging to the above mentioned genera, and exhibits a particular combination of predominant species that is distinct from that found in other individuals. There is considerable variation in the composition of the fecal microbiota among individuals (inter-individual diversity). However, studies in a single subject showed that the microbiota is quite stable over prolonged periods of time (intra-individual stability). There may be transient fluctuations in the composition of the individual microbiota under certain circumstances, for instance, associated with acute diarrheal illnesses, antibiotic therapy or to lesser extent induced by dietary interventions, but the microbiota tend to return to their typical compositional pattern. This phenomenon is called resilience. Simon and Gorbach (1984) suggested that the complexity of the intestinal ecosystem makes it most resistant to change and allows the microbiota to resume its original composition shortly after an environmental insult. On the basis of such observations, investigators suggested that microbial communities in the gastrointestinal tract contain "autochthonous" and "allochthonous" members (Savage 1977). These categories provide a useful framework for understanding patterns of diversity and stability in the communities. Autochthonous members occupy physical niches and form stable populations over long periods. In contrast, allochthonous microbes lack specific niches and are transient colonizers of the gastrointestinal tract.

However, a number of bacteria that are observed by direct microscopic examination of diluted fecal specimens cannot be grown in culture media. Prokaryotic organisms need biodiversity for growth. Thus, some 60% to 80% of the total microscopic counts are not recoverable by culture (Suau et al., 1999). Molecular biological procedures have been developed to investigate microbial composition without use of cultures. In addition, conventional methods that rely on isolation of microbes in pure culture cannot address ecological questions (Frank & Pace, 2008). Understanding of complex microbial communities requires a global approach because of the dependencies and interactions within community members. For these reasons, the use of molecular–biological technology has expanded dramatically in recent years and now permits the study of complex microbial communities in the field of environmental microbiology (natural microbial communities in soil, water reservoirs, oceans, etc.) and in symbiotic systems such as the human gastrointestinal tract. Interestingly, results of an initial analysis of bacterial genes in human feces revealed that most DNA sequences corresponded to previously undescribed microorganisms (Suau et al., 1999). The novel genomic approach thus provides new insights into the microbial communities of the human gut.

3.1 MOLECULAR TECHNIQUES

Molecular biological techniques are based on the differences in the sequence of nucleotides in the microbial genes, so that nucleotide sequences can be used as a fingerprint to identify genes, functions and species. A particularly powerful approach, labeled "metagenomics," is described as the study of genetic material recovered directly from environmental samples bypassing the need to isolate and culture individual bacterial community members. The metagenome is the collective genetic content of the combined genomes of the constituents of an ecological community (Frank & Pace, 2008). The microbiome is defined as the collective genome of the microbial symbionts in a host animal (Turnbaugh et al., 2007). Through this methodology, the genomes of entire populations of microorganisms are studied in parallel. The starting material for a metagenomic study is a mixture of DNA from a community of cells that may include bacterial, archaeal, eukaryotic and viral species at different levels of diversity and abundance. These studies can generate a profile of the kinds of organisms that inhabit an environment and delineate the genetic and metabolic capacities of the entire community (who is in there and what are they doing). This approach combines the power of genomics and bioinformatics, i.e., the computational tools developed to analyze large datasets of DNA sequences. The full potential of the molecular approach includes meta-transcriptomic, proteomic and metabolomic analysis of complex communities (Figure 13) (Zoetendal et al., 2008).

The first step in the study of an environmental sample is the characterization of its composition. The gold standard for molecular identification of microbial species is the phylogenetic analysis

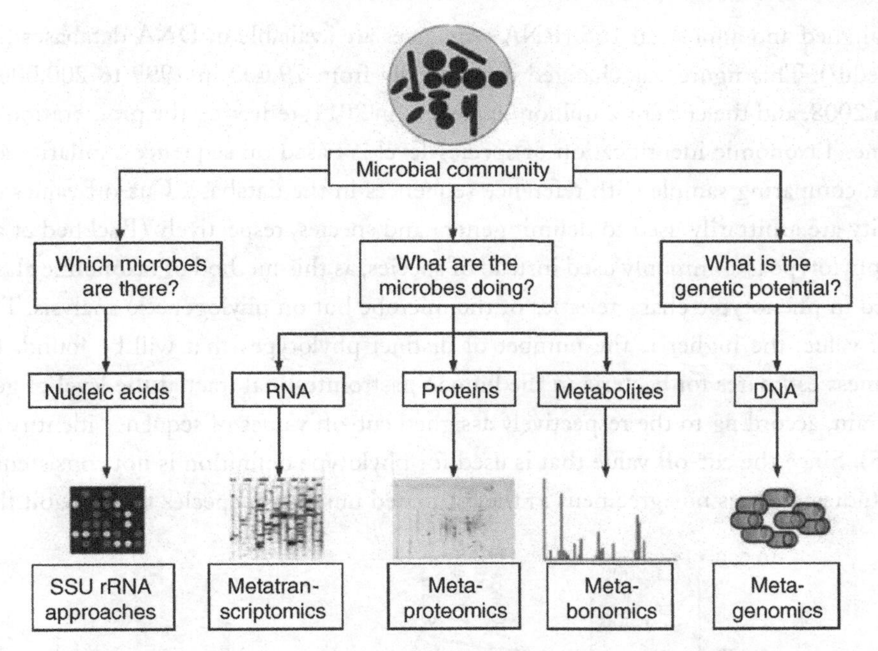

FIGURE 13: Investigation of complex microbial communities requires a global approach because of the dependencies and interactions within community members. Different "omics" technologies can be applied to profile the kinds of organisms in a community and delineate the genetic and metabolic capacities of the entire community (SSU rRNA, small sub unit of ribosomal RNA genes). Source: Figure 3 in Zoetendal EG, Rajilic-Stojanovic M, de Vos WM. High-throughput diversity and functionality analysis of the gastrointestinal tract microbiota. *Gut* 2008;57(11): 1605–15. Reproduced with permission of BMJ Publishing Group Ltd.

of genes highly conserved in all bacteria and archaea: the ribosomal RNA genes. In particular, the small subunit ribosomal RNA gene (16S rRNA) has become the standard in prokaryotes for determination of phylogenetic relationships, assessment of diversity in the environment and detection and quantification of specific populations. While there is a homologous gene in eukaryotes (18S rRNA), it is distinct, thereby rendering 16S rRNA a specific tool for extracting and identifying prokaryotes as separate from plant, animal or fungal DNA within complex samples. The gene is universally distributed among prokaryotes, allowing the comparison of phylogenetic relationships among all extant organisms and thus the construction of a "tree of life" (Pace, 2009). The 16S rRNA contains conserved and variable regions that allow taxonomic identification ranging from the domain and phylum level to the species level.

For all these reasons, analysis of 16S rRNA sequences is often the first step, and provides relatively rapid and cost-effective assessment of bacterial diversity and abundance. Currently, around

2 million aligned and annotated 16S rRNA sequences are available in DNA databases (http://rdp.cme.msu.edu/). This figure has changed dramatically from 79,000 in 1999 to 200,000 in 2004, 600,000 in 2008, and the current 2 million sequences in 2011, reflecting the proliferation of studies on this gene. Taxonomic identification of species-level is based on sequence similarity analysis of 16S rRNA, comparing sample with reference sequences in the database. Cut-off values of 95% or 98% identity are arbitrarily used to delimit genera and species, respectively (Backhed et al., 2005). The term phylotype is commonly used instead of species, as this method of taxonomic classification is not based in phenotypic characteristics of the microbe but on phylogenetic analysis. The higher the cut-off value, the higher is the number of distinct phylotypes that will be found. Figure 14 shows richness estimates for bacteria in the human gastrointestinal tract at the level of genus, species and strain, according to the respectively assigned cut-off values of sequence identity (Backhed et al., 2005). Since the cut-off value that is used for phylotype definition is not consistent between different studies, there is no agreement in the estimated number of species that inhabit the human

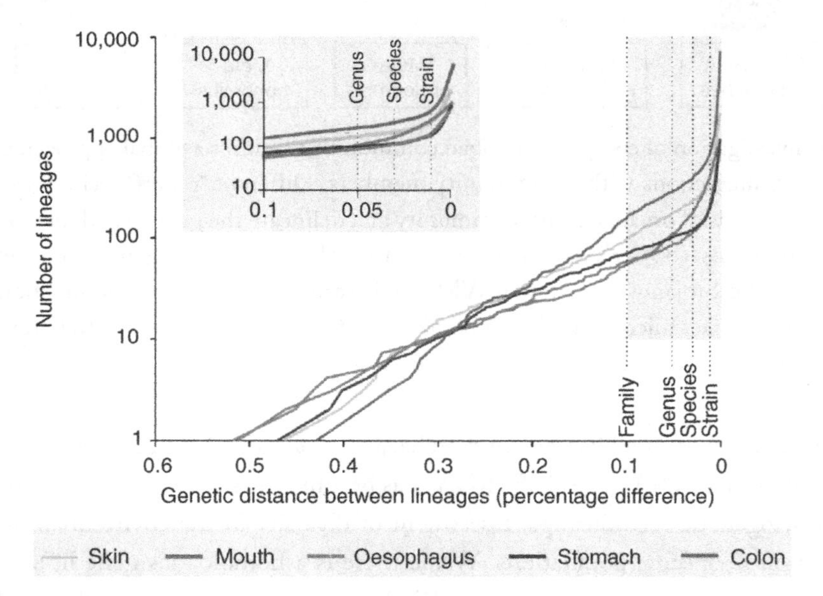

FIGURE 14: Differences in nucleotide sequence of 16S rRNA gene are used for taxonomic identification and clustering. The graph shows numbers of different 16S rRNA sequences (lineages) for bacteria in various human body sites (skin, mouth, esophagus, stomach and colon) at different levels of 16S rRNA sequence identity (genetic distance). Percentage difference in sequences at 5% (95% identity) is arbitrarily used to identify members from the same genus, whereas 3% difference (97% identity) is used for species level. The inset depicts a portion of the same data at a larger scale. Source: Figure 2 in: Dethlefsen L, McFall-Ngai M, Relman DA. An ecological and evolutionary perspective on human-microbe mutualism and disease. *Nature* 2007 Oct 18;449(7164): 811–8. Reproduced with permission of Nature Publishing Group.

gastrointestinal tract with proposed figures ranging from hundreds to thousand species (Eckburg et al., 2005; Ley et al., 2006a; Frank & Pace, 2008).

Different 16S rRNA-based techniques are appropriate for different investigations, including identifying microbial species or phylotypes, quantifying microbial taxa or making broad comparisons of microbial communities. Table 2 summarizes the most common techniques based on this

TABLE 2: 16S rRNA methods of analyzing microbial communities.

METHOD	MAIN USE	ADVANTAGES	LIMITATIONS
Oligonucleotide hybridization (FISH, flow cytometry, microarrays)	Detect and quantify known phylogenetic groups	Can be high-throughput; can reveal spatial relationships; phylogenetic identification of visible cells	Detects only taxa that hybridize to the chosen probes
Community profiling (DGGE, tRFLP)	Compare communities	Rapid, inexpensive assessment of abundant 16S rRNA sequence variants	Broad-range PCR bias; additional work needed to identify groups represented in profiles; hard to compare analyses done at different times
quantitative PCR	Detect and quantify known phylogenetic groups	Rapid; high throughput	Detects only taxa targeted by the chosen probes and primers
16S rRNA sequencing	Phylogenetic identification of microbes; generates data for other 16S rRNA-based methods	Identification to strain level; can detect novel taxa; analysis possible at multiple phylogenetic levels	Broad-range PCR bias; expensive; laborious data analysis

FISH, fluorescent in situ hybridization; DGGE, denaturing gradient gel electrophoresis; tRFLP, restriction fragment length polymorphism; PCR, polymerase chain reaction. The broad-range PCR might not include all taxa or accurately represent their abundance. Adapted from Dethlefesen et al. (2006).

gene. The identification of a whole bacterial population as well as a single bacterial species or strain requires different technical strategies. The highly conserved regions of the 16S rRNA gene are used to design universal primers or probes capable to detect all members of the environmental sample, while the variable regions are used to design more specific primers or probes that target a single bacterial genus, species or strain, depending on the selectivity of the targeted sequence.

Ultimately, the most powerful molecular tools rely in shotgun whole genome sequencing. The 16S rRNA approach may be useful for deciphering taxonomic diversity of members within the microbial community in a particular niche but cannot tell about the total gene content and biological functions present in the community. The decreasing cost and increasing speed of DNA sequencing, coupled with the advances in computational analyses of large datasets, have made it feasible to analyze complex mixtures of entire genomes with reasonable coverage. The resulting information describes the collective genetic content of the community from which functional and metabolic networks can be inferred. This approach is currently being used by large-scale research projects on the human microbiome in different parts of the World (Turnbaugh et al., 2007; Mullard, 2008). With the advent of methods for large-scale and high-throughput genotyping, the generation of data is growing in a logarithmic way, beyond the capacity of analysis that the data producers can accomplish themselves. In addition, shared data sets are much larger, richer and of higher quality than individual laboratories could normally generate. For these reasons, the International Human Microbiome Consortium (IHMC) was constituted in Heidelberg in October 2008 by an agreement between research funding agencies and investigators integrated in large-scale research projects from different countries. The IHMC is a global effort to characterize the role of the human microbiome in the maintenance of health and in disease. The IHMC main objectives are to coordinate the activities and policies of the international groups studying the human microbiome and to promote the generation of a robust data resource that is freely available to the global scientific community (http://www.human-microbiome.org/).

3.2 DIVERSITY, STABILITY AND RESILIENCE

Molecular studies based on 16S rRNA gene sequencing have highlighted that only 7 to 9 of the 55 known divisions or phyla of the domain Bacteria are detected in fecal or mucosal samples from the human gut (Eckburg et al., 2005; Backhed et al., 2005; Ley et al., 2006a; Gill et al., 2006; Franck & Pace, 2008). Moreover, such studies also revealed that more than 90% of all the phylotypes belong to just 2 divisions: the Bacteroidetes and the Firmicutes (Figure 15). The other divisions that have been consistently found in samples from the human distal gut are Proteobacteria, Actinobacteria, Fusobacteria and Verrucomicrobia (Zoetendal et al., 2008). Of the 13 divisions of the domain Archea, only one species (*Methanobrevibacter smithii*) seems to be represented in the human distal gut

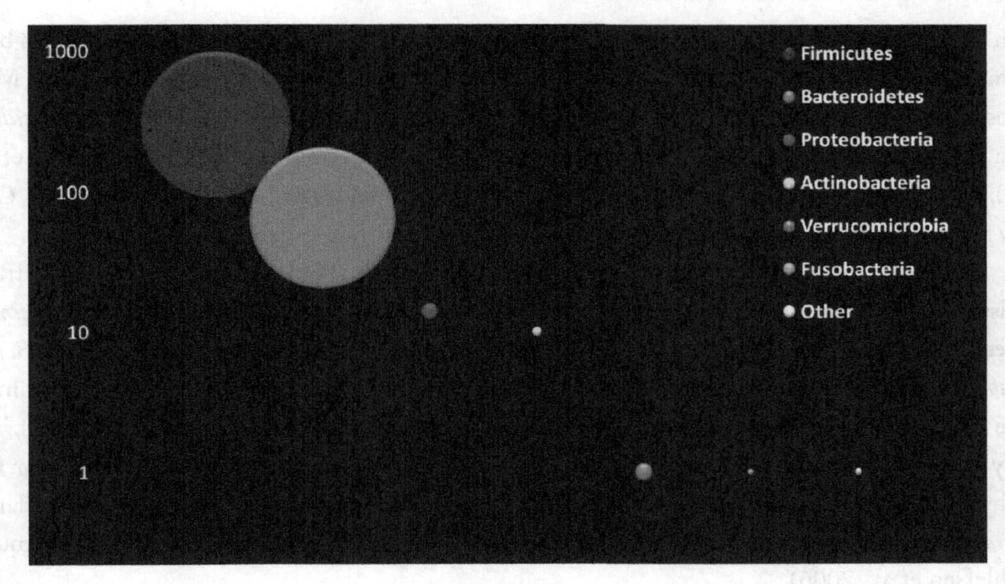

FIGURE 15: Composition of the human enteric microbiota as determined by 16S rRNA sequencing of fecal and mucosal samples from the gastrointestinal tract. Only 7 to 9 of the 55 known divisions or phyla of the domain Bacteria are detected. The graphic shows number of phylotypes (species) of each bacterial division, and the size of the dotes is proportional to the number of different strains in each division. More than 90% of all the phylotypes belong to 2 divisions: the Bacteroidetes and the Firmicutes (data from the study by Eckburg et al (2005). Source: drawing by the author using data published by Eckburg et al. 2005.

microbiota (Eckburg et al., 2005). Thus, at the division level, the human intestinal ecosystem is less diverse than other ecosystems on Earth, like soils and ocean waters which may contain 20 or more divisions (Backhed et al., 2005). Interestingly, 62% of the phylotypes detected in the study by Eckburg et al. (2005) were novel "species," i.e., never sequenced before that time, and 80% represented sequences from species that had not been cultivated.

The Firmicutes division is usually the most numerous and diverse in the human distal gut. The *Clostridium* cluster XIVa is a family-level taxon within the Firmicutes containing many butyrate-producing strains, including *Roseburia* and relatives that can degrade starch and inulin (Flint et al., 2008). A second abundant group within this cluster is related to the genus *Eubacterium*, and ferments lactate and acetate to butyrate and hydrogen. Important non-butyrate producing members of cluster XIVa are *Ruminococcus torques* and *R. gnavus*, which are among the primary mucin-degrading organisms (Dethlefsen et al., 2006). The *Clostridium* cluster IV is another prominent family-level taxon within the Firmicutes containing *Faecalibacter prausnitzii*, abundant fermenter of starch and inulin to butyrate and lactate. Reduction in relative abundance of *F. prausnitzii* has been incriminated in the pathogenesis of chronic inflammatory bowel diseases (Sokol et al., 2008). Most interestingly, recent findings suggest that members of clusters IV and XIVa of genus *Clostridium* play a major role in the induction of regulatory T cells in gut lymphoid follicles (Atarashi et al., 2011). Other groups of Firmicutes given the genus names *Peptococcus, Peptostreptococcus* and *Clostridia* are the predominant proteolytic and amino acid-fermenting organisms in the colon.

Bacteroides, of the Bacteroidetes division, are prominent starch degraders and many strains are also capable of degrading some types of structural polysaccharides. Most common *Bacteroides* species are *B. vulgatus* (31% of Bacteroidetes sequences in the study by Eckburg et al., 2005), *B. thetaiotaomicron* (13% of sequences) and *B. fragilis*. The role of *B. thetaiotaomicron* in the development of the intestinal epithelium has been extensively studied in recent years (Hooper et al., 2001). The ability of some *Bacteroides* species to import oligosaccharides into their periplasmic space for further hydrolysis (thus monopolizing the hydrolysis products) might contribute to their abundance. *Bacteroides* are also primarily responsible for removing the sulfate ester-linked substituents of mucin (Dethlefsen et al., 2006).

In the Actinobacteria division, *Bifidobacterium* species are common and well known colonizers of the human colon. The genus mainly includes inulin and starch degraders, and some strains that are mucin degraders. Lactate is the primary fermentation product of bifidobacteria, much of which is converted to butyrate by secondary fermenters. Sulfate-reducing bacteria are found in five distinct genera in the Delta subdivision of the Proteobacteria phylum. Hydrogen-consuming, sulfate-reducing bacteria are found in two of these genera; the remaining genera consume partially reduced fermentation products (e.g., lactate) while reducing sulfate to sulfide (Dethlefsen et al., 2006). The Enterobacteriacea (*Escherichia, Enterobacter, Klebsiella, Proteus*, etc.) belong to the Gamma-Proteobacteria subdivision, and they are found at very low level of abundance. Among the Verrucomicrobia division, only one species is commonly detected, namely, *Akkermansia muciniphila*, a newly described mucin-degrading Gram-negative anaerobe (Derrien et al., 2004).

While the large majority of bacteria identified in samples from the human gut belong to just two divisions, there is huge diversity between individuals at lower taxonomic level. In the dataset of 13,335 16S rRNA sequences pooled from three individuals, over half of them were present in only one subject (Eckburg et al., 2005). In another study including samples from 14 individuals, the resulting data set of 16S rRNA sequences revealed that most (70%) of the identified species-level phylogenetic types (phylotypes) were unique to each person (Ley et al., 2006b). These observations are representative of the great inter-individual differences. At the level of species and strains, microbial diversity between individuals is highly remarkable so that each individual harbors his or her own distinctive pattern of bacterial composition (Ley et al., 2006a; Turnbaugh et al., 2009). This pattern appears to be determined partly by the host genotype (Zoetendal et al., 2001) and by initial colonization at birth via vertical transmission (Ley et al., 2006a).

The differences between individuals are greater than the differences between different sampling times in the same individual (intra-individual stability). Fecal 16S rRNA sequences from 14 adults over the course of a year showed that community structure in each host was generally stable during this period (Ley et al., 2006b). Another interesting study surveyed bacteria from different body sites, including gut (stools), oral cavity, ear, nostrils, hair and skin surfaces, in 9 healthy adults on four occasions. It was found that community composition was determined primarily by body habitat, with large differences form skin to mouth or gut. Within habitats, however, individuals exhibited minimal temporal variability (Costello et al., 2009). Intra-individual differences between different sampling sites of the distal gastrointestinal tract are also minimal. In the colon, bacterial composition in the lumen varies from cecum to rectum. Enterobacteriacea, lactobacilli and bifidobacteria are more abundant in the cecum, whereas bacteroidales and clostridia are more abundant in the rectum (Marteau et al., 2001). However, the community of mucosa-associated bacteria is highly stable from terminal ileum to the large bowel in a given individual (Zoetendal et al., 2002; Lepage et al., 2005; Eckburg et al., 2005).

In healthy adults, the fecal composition is stable over time, but temporal fluctuations due to ordinary environmental factors (foods, travelling, etc.) can be detected and may involve up to 20% of the dominant groups as assessed by DGGE (Zoetendal et al., 2001). In some circumstances, changes due to a stronger environmental insult can be more striking, but are transitory. For instance, the microbiota profiles in acute diarrhea exhibit decreased diversity and, in some cases, overgrowth of selected bacteria or bacterial groups (Mai et al., 2006). In healthy adults, oral treatment with the antibiotic ciprofloxacin for 5 days influenced the abundance of about a third of the bacterial taxa in fecal samples, decreasing the taxonomic richness, diversity and evenness of the community. The taxonomic composition of the community closely resembled its pretreatment state by 4 weeks after the end of treatment, but several taxa failed to recover within 6 months (Dethlefsen et al., 2008).

The rapid return to the original community structure after a strong insult is indicative of factors promoting community resilience. The human distal gut microbiota is composed of a consortium of species that are specific to the host and resistant to modifications over time. This suggests that autochthonous microbes are the main colonizers of the large intestine, and allochthonous microbes are outnumbered by autochthonous microbes. This is contrasting with the communities in the upper gastrointestinal tract (stomach and small intestine) where the impact of allochtonous microbes (bacteria in transit) is remarkable and community composition is variable (Walter & Ley, 2011).

3.3 THE HUMAN GUT MICROBIOME AND THE ENTEROTYPES

For a better understanding of the functional impact of gut microbes on human health, it is crucial to assess their genetic potential. This is currently being investigated by whole genome shotgun sequencing studies of total DNA extracted from samples and subsequent assembly of sequences to predict genes through bioinformatic analysis. Since crude DNA extracts are being used, the procedure avoids the previous step of broad-range PCR, which commonly precedes 16S rRNA analysis. The so-called universal primers used for PCR amplification may miss certain species or reflect inaccurately their abundance. The metagenomic procedures have higher definition power (all genes are being analyzed) and more accuracy than 16S rRNA sequencing but less sensitivity for the detection of under-represented species.

In a cohort of 124 European adult subjects, a total of 3.3 million microbial genes were characterized by metagenomic analysis of fecal samples (Qin et al., 2010). This effort has provided the first gene catalogue of the human gut microbiome (Table 3). Each individual carries an average of 600,000 non-redundant microbial genes in the gastrointestinal tract. This figure suggests that most of the 3.3 million genes in the catalogue are shared. It was found that around 300,000 microbial genes are common in that they are present in at least 50% of individuals. On the other hand, a large pool of 2.4 million genes is considered rare since they are only found in few individuals.

The microbial gene catalogue was aligned with the known genomes of bacterial and archeal species that have been fully sequenced and are available in public databases. This procedure aimed at exploring the presence of known microbial species in the cohort of individuals. It was found that 99% of genes in the catalogue are bacterial, and that the entire cohort of individuals harbors between 1000 and 1150 prevalent bacterial species, with at least 160 species per individual (Qin et al., 2010). This figure differs from other studies using 16S rRNA methods, which have proposed higher diversity of species in each individual. As mentioned, 16S rRNA methods can be more sensitive than whole genome sequencing due to the broad range PCR step, but probably the metagenomic data reflect more accurately the dominant species in the human gut. A further study by the same group included data from Japanese and American individuals and estimated abundance of bacteria at the level of phylum and genus using the metagenomic approach (Figure 16). The study confirmed

TABLE 3: The human gut microbiome.

MICROBIAL GENES IN THE HUMAN GUT	NUMBER OF GENES
Non-redundant gene set in the cohort	3,299,822
Median gene set per individual	590,384
Common genes (present in at least 50% of individuals)	294,110
Rare genes (present in less than 20% of individuals)	2,375,655

Data from Qin et al. (2011).

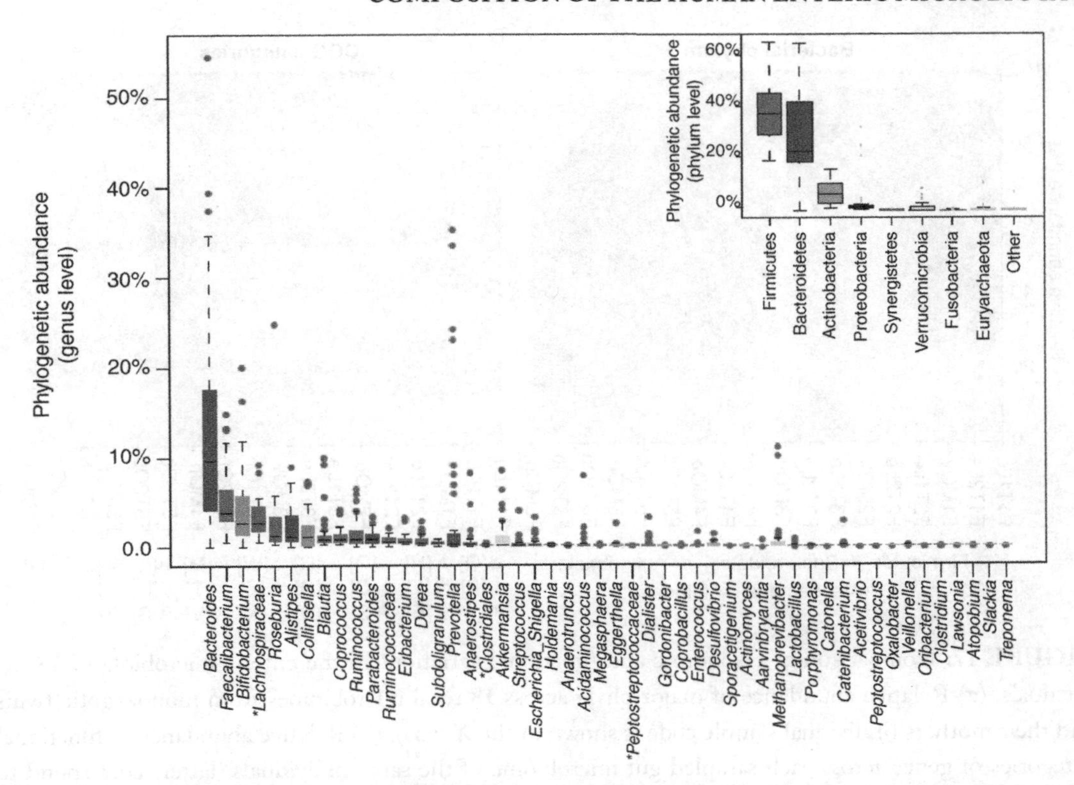

FIGURE 16: Genus abundance variation box plot for the 30 most abundant genera as determined by metagenomic sequencing of human fecal samples. Genera are colored by their respective phylum (see inset for color key). Inset shows phylum abundance box plot. Genus and phylum level abundances were measured using reference-genome-based-mapping. Source: Figure 1b in Arumugam M et al. Enterotypes of the human gut microbiome. *Nature* 2011 May 12;473(7346): 174–80. Reproduced with permission of Nature Publishing Group.

that the Firmicutes and Bacteroidetes divisions (phyla) constitute the vast majority of the dominant gut microbiota. Interestingly, *Bacteroides*, *Faecalibacterium* and *Bifidobacterium* are the most abundant genera but their relative abundance is highly variable across samples (Arumugam et al., 2011). The most striking difference with culture-based studies performed in the past (Simon & Gorbach, 1984) is the presence of *Faecalibacterium* in the top list. Indeed, this obligate anaerobe is very difficult to isolate in culture.

The functional assignments of the gene catalogue are being investigated. There are about 20,000 biological functions coded by these microbial genes, and a third of them are already well characterized (Qin et al., 2010). Interestingly, a metagenomic study in 18 American individuals has shown that diversity in taxonomy does not correlate with diversity at the gene functional level (Turnbaugh et al., 2009). This observation is illustrated in Figure 17, whereas taxonomic profiles

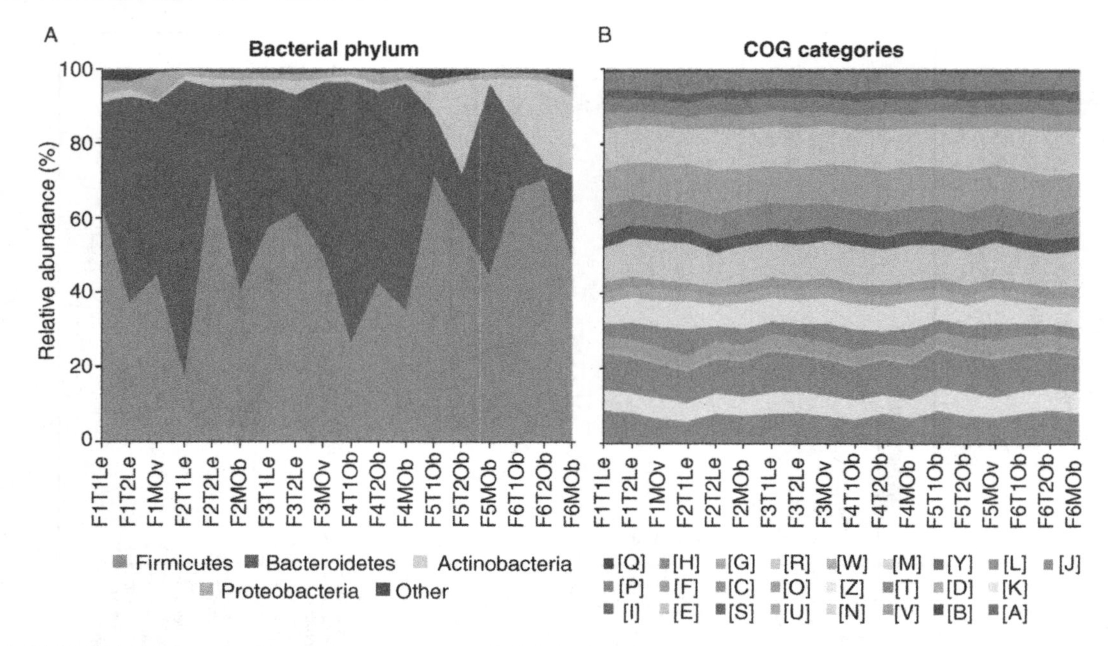

FIGURE 17: Comparison of taxonomic and functional variations in the enteric microbiota of 18 individuals. (a) Relative abundance of major phyla across 18 fecal microbiomes from monozygotic twins and their mothers (individual sample code is shown in the *X*-axis). (b) Relative abundance of functional categories of genes across each sampled gut microbiome of the same individuals (letters correspond to categories in the Clusters of Orthologous Groups of proteins database). Source: Figure 3 in: Turnbaugh PJ, Hamady M, Yatsunenko T, Cantarel BL, Duncan A, et al. A core gut microbiome in obese and lean twins. *Nature* 2009: 457: 480–84. Reproduced with permission of Nature Publishing Group.

show high variation across individuals, relative abundance of categorized gene functions is fairly stable. Redundancy of functions at the community level (genes shared among different species) may explain why the enteric microbiota is better defined by the collective genome rather than by taxonomic assignment of the community members.

An interesting observation of the metagenomic studies is the interactions or correlations between clusters of species within the enteric microbial community. A complex pattern of species relatedness emerged from the network analysis where pairwise correlations of species abundance were tested across the cohort of 124 individuals (Qin et al., 2010). Interestingly, the network analysis yielded a number of positive or negative correlations of abundance of the most prevalent species. This observation suggested that within the community, some species interact positively with each other, and their abundances fluctuate up and down together across different individuals' microbiota.

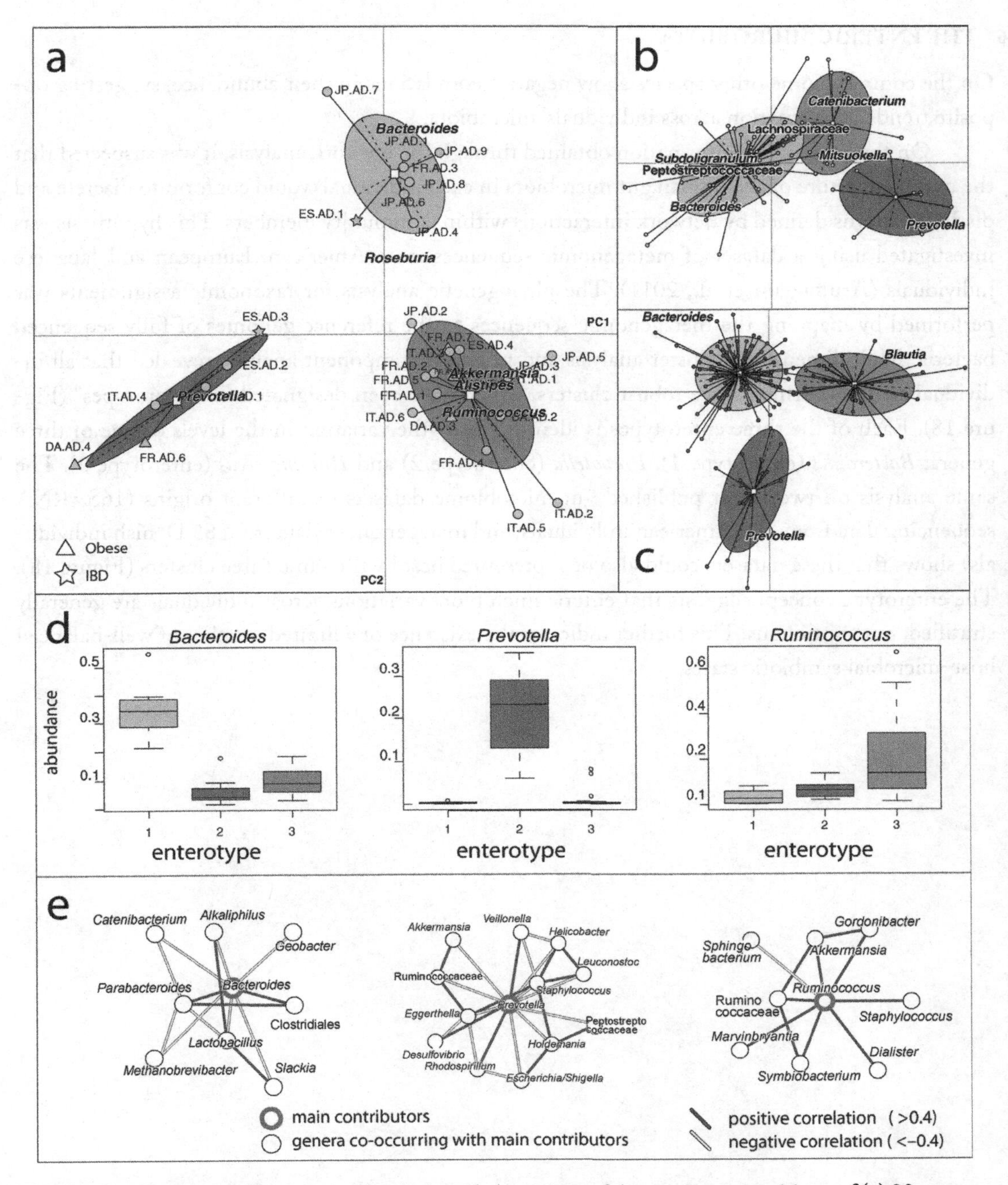

FIGURE 18: Principal Component Analysis and clustering, of the genus compositions of (a) 33 metagenomes from American, European and Japanese individuals, (b) Danish subset containing 85 metagenomes and (c) 154 16S rRNA sequenced samples reveal three robust clusters that are called enterotypes. (d) Abundances of the main contributors of each enterotype. (e) Co-occurrence networks of the three enterotypes showing positive and negative correlations between genera. Source: Figure 2 in Arumugam M et al. Enterotypes of the human gut microbiome. *Nature* 2011 May 12;473(7346): 174–80. Reproduced with permission of Nature Publishing Group.

On the contrary, some other species show negative correlations in their abundance, suggesting opposite trends in fluctuation across individuals' microbiota.

On the basis of the information obtained through the network analysis, it was suspected that the overall structure of the human gut microbiota in each individual would conform to discrete and distinct patterns defined by network interactions within community members. This hypothesis was investigated using a dataset of metagenomic sequences from American, European and Japanese individuals (Arumugam et al., 2011). The phylogenetic analysis for taxonomic assignments was performed by mapping the metagenomic sequences to the reference genomes of fully sequenced bacteria. Multidimensional cluster analysis and principal component analysis revealed that all individual samples formed three robust clusters, which have been designated as "enterotypes" (Figure 18). Each of the three enterotypes is identifiable by the variation in the levels of one of three genera: *Bacteroides* (enterotype 1), *Prevotella* (enterotype 2) and *Ruminococcus* (enterotype 3). The same analysis on two larger published gut microbiome datasets of different origins (16S rRNA sequencing data from 154 American individuals, and metagenomics data from 85 Danish individuals) shows that these datasets could also be represented best by the same three clusters (Figure 18). The enterotype concept suggests that enteric microbiota variations across individuals are generally stratified, not continuous. This further indicates the existence of a limited number of well-balanced host–microbial symbiotic states.

• • • •

CHAPTER 4

Acquisition of the Enteric Microbiota

Humans are born with a sterile gastrointestinal tract that is successively colonized with microbial populations until adult-like communities stabilize (Walter & Ley, 2011). The birth canal is heavily populated by microbes, suggesting that the vagina has evolved to provide the primary inoculum (Dominguez-Bello et al., 2011). The human vagina harbors a microbial ecosystem with relatively few predominant species which are site-specific. At the time of delivery, the human vagina is dominated by *Lactobacillus* and *Prevotella* species. Recent 16S rRNA sequencing studies have shown that vaginally delivered babies acquire at birth their own mother's vaginal bacteria (Dominguez-Bello et al., 2010). These bacteria can be found in the skin and mouth of the baby and are present in the first meconium. Different body sites of the newborn are colonized with essentially the same microbiota that was inherited vertically from their mothers and only later develop the distinct microbial communities found at these sites in adults. However, in contrast to vaginally delivered babies, cesarean section babies harbor bacterial communities that resemble those of the skin, comprising *Staphylococcus*, *Corynebacterium*, and *Propionibacterium* species (Figure 19). These communities in the baby are not closer to that of the mother's skin than to the skin of other women, indicating that the initial inoculum in these babies is provided by other people with whom they are in contact. Human-associated bacteria are common in hospital environments, and it has been shown that similarity in fecal bacteria composition among hospitalized neonates is high and increases with days of hospital stay, revealing the acquisition of bacterial communities from the hospital environment (Scwiertz et al., 2003).

After the natural primary inoculation at birth, infants have multiple exposures to human microbes. The days and weeks following birth mark a period of extraordinary transformation in the microbial ecology of the human gut. This process has been described as chaotic because of its model of progression fits better with a stochastic rather than a deterministic system (Walter & Ley, 2011). The strains acquired from mothers' vagina and skin are replaced by other strains of less certain origin. The newborn's enteric microbiota has relatively few species and lineages, but diversity increases rapidly. Culture-based studies suggested that facultative anaerobes establish first and reduce the environment so that strict anaerobes can be established in sequence. Consistent with such studies, 16S

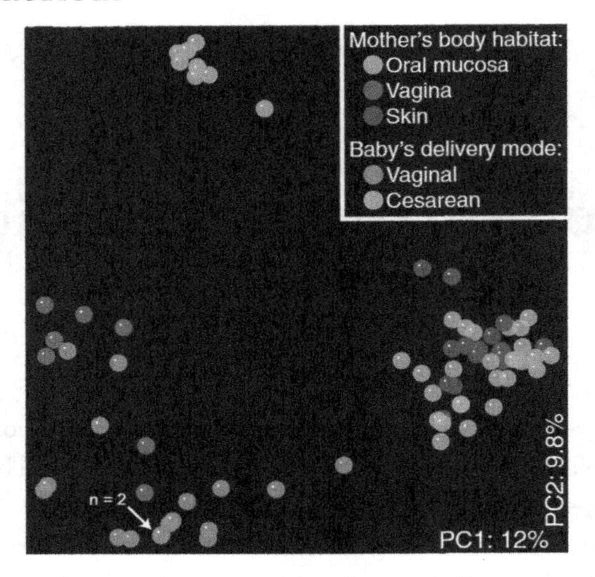

FIGURE 19: The first microbiotas of human newborns are primarily structured by delivery mode. Data points in the graph represent the bacterial communities in each individual sample as determined by 16S rRNA sequences analysis. The data points are clustered according to similarity of the bacterial community structure by principal coordinates analysis. Each point is colored according to the mother's body habitat or the newborn's delivery mode. Source: Figure 1A in Domínguez-Bello MG, Costello EK, Contreras M, et al. Delivery mode shapes the acquisition and structure of the initial microbiota across multiple body habitats in newborns. *Proc Natl Acad Sci U S A* 2010;107: 11971–11975. Reproduced with permission of PNAS USA.

rRNA methodologies have shown that the earliest colonizers are often facultative anaerobes organisms (e.g., *Staphylococcus*, *Streptococcus* and *Enterobacteriaceae*), whereas the later colonizers tend to be strict anaerobes (*Eubacterium* and *Clostridium*) (Palmer et al., 2007). The *Bacteroides* varies greatly from baby to baby in the timing of their first appearance, but they are consistently present in nearly all babies by 1 year of age. Only a few of the infants that have been investigated keep evident similarities in composition to mothers' breast milk, vaginal or stool samples. Another striking feature of the infant gut microbiota is the tremendous temporal variations in dominant microbial populations within the same individual over the course of weeks and months following birth. By the end of the first year of life, however, microbial composition of fecal samples becomes more stable and increasingly reflects a typical adult-like microbial community with predominance of Firmicutes and Bacteroidetes. Commonly shared temporal patterns of colonization during this year or the events that shape patterns of colonization have not been identified.

Factors such as host genetic background and diet may be important determinants of the individual microbiota composition. Zoetendal et al. (2001), using DGGE, suggested that host genotype affects the composition of the fecal microbiota. In that study, the authors examined fecal samples from 50 donors of varying relatedness. Fingerprint profiles from monozygotic twins showed significantly higher similarity than those from unrelated individuals. In addition, a positive relationship between the similarity indices and the genetic relatedness of the hosts was observed. In contrast, profiles of marital partners, which are living in the same environment and which have comparable feeding habits, showed low similarity which was not significantly different from that of unrelated individuals. However, the study could not differentiate effects of genetic relatedness from those of early cohabitation, i.e., environmental exposure during the first stage of life. This issue was addressed in a study including 31 monozygotic and 23 dizygotic twin pairs (Turnbaugh et al., 2009). Analysis of 16S rRNA sequences in fecal samples revealed that individuals from the same family had a more similar bacterial community structure than unrelated individuals. However, there was no significant difference in the degree of similarity in the gut microbiotas of monozygotic compared with dizygotic twin pairs. This observation highlights the impact of early environmental events on microbiota composition. In an experiment with genetically modified mice, the composition of the enteric microbiota was dictated by the surrogate mother regardless of the genotype of their offsprings (Ley et al., 2006). These experimental conditions suggested that early colonizers rather than host genes are key factors influencing community assembly. In addition, the data reflect that the consortia of bacterial taxa are not random, at any given time point, indicating that the microbes depend on each other within the consortium (Dominguez-Bello et al., 2011).

Changes in diet have been shown to alter the microbiota composition (Ley et al., 2006b; Wu et al., 2011). In a recent study in adults, the *Bacteroides* enterotype was associated with dietary habits consisting predominantly of animal protein and saturated fats, suggesting that meat consumption as in a Western diet characterized this enterotype. The *Prevotella* enterotype, in contrast, was associated with diets with high proportion of carbohydrates and simple sugars, indicating association with a carbohydrate-based diet more typical of agrarian societies. These observations suggest that dietary factors shape the enteric microbiota, but no such data exist from infants during the first year of life.

Taken together, the successional mechanisms involved in the development of the enteric microbiota from founder communities to the adult communities appear to be dependent on environmental exposures and modulated by multiple endogenous factors including host-specific, niche-specific and community-specific characteristics.

4.1 HYGIENE AND IMMUNE DYSREGULATION

Atopic diseases, such as allergies and asthma, and chronic idiopathic inflammatory bowel diseases, like ulcerative colitis and Crohn's disease, show higher incidence in children born by cesarean-section than in those born by vaginal delivery (Bager et al., 2008; Bager et al., 2011; Van Nimwegen et al., 2011). As mentioned, birth by cesarean section has an impact on the profile of microbes acquired by the baby and may affect the development of the microbial ecosystem in the gastrointestinal tract (Dominguez-Bello et al., 2011). Multiple factors in modern life, such as sanitation, urban living, antibiotic use, etc., can influence microbial exposure and gut colonization. Altered microbial colonization might be in the origin of a defective instruction of the immune system and an increased incidence of immune-regulatory disorders (Van Nimwegen et al., 2011; Bernstein & Shanahan, 2008).

The incidence of allergic disorders in North America and Western Europe increased from the late 19th century and appears to have doubled in some decades, particularly during the 1960s and 1970s. A link between increasing allergies and the modern hygienic lifestyle was initially suggested, giving rise to the so-called Hygiene Hypothesis (Strachan et al., 1996). Allergic disorders depend mainly on immune responses by Th2 lymphocytes. Therefore, the original assumption was that "hygiene" provided diminished exposure to infections by microorganisms driving Th1 activity that should crossregulate Th2 activity (Romagnani, 1994).

Such Th1/Th2 imbalance, however, cannot explain the simultaneously increased incidence of several other immunological disorders like Crohn's disease, type 1 diabetes (insulin-dependent) and multiple sclerosis, which all are primarily driven by Th1 cells. Epidemiological data demonstrate a steady increase of various immunoregulatory disorders in westernized countries during the past several decades (Figure 20). Notably, the incidences of allergic disorders (Th2) and type 1 diabetes (Th1) correlates closely both within Europe and North America (Bach, 2002). Asthma became a

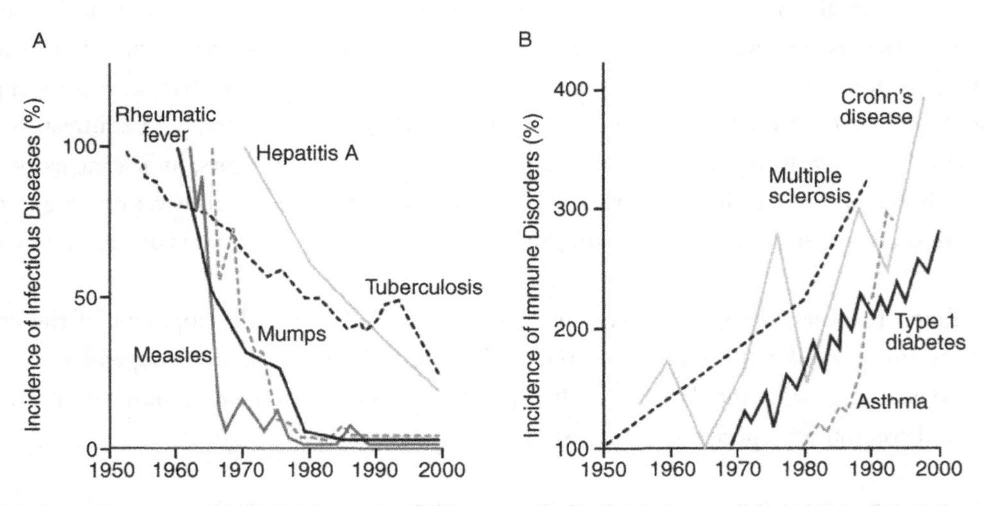

FIGURE 20: Inverse relation between the incidence of infectious diseases (Panel A) and the incidence of immune disorders (Panel B) from 1950 to 2000. Data are derived from reports in US and European countries. Source: Figure 1 in: Bach JF. The effect of infections on susceptibility to autoimmune and allergic diseases. *N Engl J Med* 2002;347: 911–920. Reproduced with permission of NEJM.

burden among children and young adults in Europe during the last quarter of the 20th century, with very high rates of prevalence (median 20.7%, from 8.5% in Pavia to 32.0% in Dublin). The incidence is now increasing in Asia and Eastern Europe (Pearce et al., 2007). In parallel, an increasing incidence of type I diabetes is also being reported in several countries (Diamond project group 2006). Multiple sclerosis has also shown an impressive increment (Bach, 2002). Inflammatory bowel diseases (IBD), including both Crohn's disease and ulcerative colitis, have been likewise on the rise during the last decades of the past century (Bernstein & Shanahan, 2008). Only environmental factors can explain the rapid changes in the trend observed over the past few decades. Interestingly, "hygiene" has repeatedly been identified by epidemiological studies as a risk factor for developing allergies, IBD, type I diabetes or multiple sclerosis (Guarner et al., 2006).

The original "hygiene hypothesis" suggested a link between the decreased burden of pathogens and the development of immunoregulatory disorders. However, neither reduced Th1-cell stimulation nor falling infection rates provide an adequate explanation for the rise of immune-mediated disorders in industrialized countries. Since these disorders are at least partly attributable to aberrant immune-regulation, authors have suggested that modern "lifestyle" might limit the normal development of regulatory immune mechanisms (Guarner et al., 2006; Bernstein & Shanahan, 2008).

A dynamic mutualism between the human host and the commensal microbiota is likely to have important implications for health. The vast numbers of bacteria that line mucosal surfaces and the skin interact directly with the host and have evolved mechanisms for co-existence over millions of years. As reviewed in Section 2.3, the decision making between induction of productive systemic-type immunity, with the potential for tissue damage and inflammation, versus a tolerogenic response appears to be largely instructed by the microbial impact on antigen-presenting cells and T cells (Figure 12). In the absence of "danger signals" derived from microbe associated molecular patterns (MAMPs), conditioned antigen-presenting cells may induce various subsets of regulatory T cells (Treg), which by their cytokines IL-10 and TGF-beta, or by direct cellular interactions, may suppress both Th1 and Th2 responses as well as innate immune activity. Such adequate balancing of the immune system appears to depend on appropriate "cross-talk" between microbes, innate immunity and adaptive responses early in the newborn period.

Microbial exposure may normally provide a natural barrier against immune-pathology by triggering homeostatic proliferation of Treg subsets. Data from animal models of autoimmunity and from patients suggest a link between defective Treg activity and immune-mediated diseases (Guarner et al., 2006). Individuals with allergic disorders have deficient Treg activity. In multiple sclerosis, putative Treg have been demonstrated in peripheral blood, but their capacity to regulate potentially disease-causing effector T cells is impaired. In healthy individuals, T cells directed against a disease-relevant islet cell auto-antigen were reported to exhibit a regulatory phenotype, but in patients with type I diabetes these T cells were pro-inflammatory Th1 subsets.

These observations emphasize a putative role of gut bacteria in the origin of the imbalance of the immune system in such conditions. The absence or relative deficiency of commensal communities able to stimulate regulatory pathways and Treg expansion might compromise immune homeostasis. Dysbiosis or abnormal composition of the gut microbiota has been investigated in some disease states, as reviewed in Chapter 5. Available data suggest that the composition of the gut microbiota is different in healthy individuals than in patients with immunoregulatory disorders such as allergies and chronic inflammatory bowel disease. Whether the observed changes are secondary or part of the pathogenic mechanism is unknown. Currently, there are no data on microbiota dysbiosis in patients with multiple sclerosis, but experimental data have shown that the intestinal microbiota profoundly impacts the balance between pro- and anti-inflammatory immune responses during the induction of autoimmune encephalomyelitis. The implication is that microbial colonization of the gastrointestinal tract may affect extra-intestinal inflammatory diseases (Lee et al., 2011).

There is ample experimental evidence showing that certain commensal bacteria can down regulate immuno-inflammatory responses (Round & Mazmanian, 2009). A variety of bacteria directly modulate cytokine networks and innate immunity by multiple mechanisms. Messenger molecules that are important for communication among bacteria (e.g., quorum sensing) may affect intestinal physiology of the host. The quorum sensing molecule 3-oxododecanoyl-L-homoserine lactone, produced by *Pseudomonas aeruginosa*, is known to suppress the production of interleukin-12 and TNF-alpha by lipopolysaccharide-stimulated macrophages (Telford et al., 1998). Some other bacteria may diminish pro-inflammatory cytokine production by inhibiting NF-kappaB activation (Kelly et al., 2004). There is now strong evidence that some bacterial preparations (including specific *Bacteroides* and *Clostridia* strains) that are therapeutically active in experimental models of chronic inflammation can induce Treg (Mazmanian et al., 2008; Atarashi et al., 2011).

4.2 THE "OLD FRIENDS" HYPOTHESIS

Because of mankind's long evolutionary association with a number of relatively harmless microorganisms, including gut commensals, saprophytic mycobacteria and helminths, they are recognized by the human innate immune system as innocuous or even treated as friends (Figure 21). Some of these "Old Friends" have been shown to induce a regulatory-type maturation of dendritic cells, facilitating their ability to stimulate Treg responses (Rook & Brunet, 2005). Therefore, rather than eliciting productive and potentially aggressive immunity, these microorganisms skew immune responses towards regulatory modulation.

The "Old Friends" are microorganisms associated with farms, untreated water supplies, cowsheds, pets, fermented foods, etc. and have been part of the human microbiological environment for

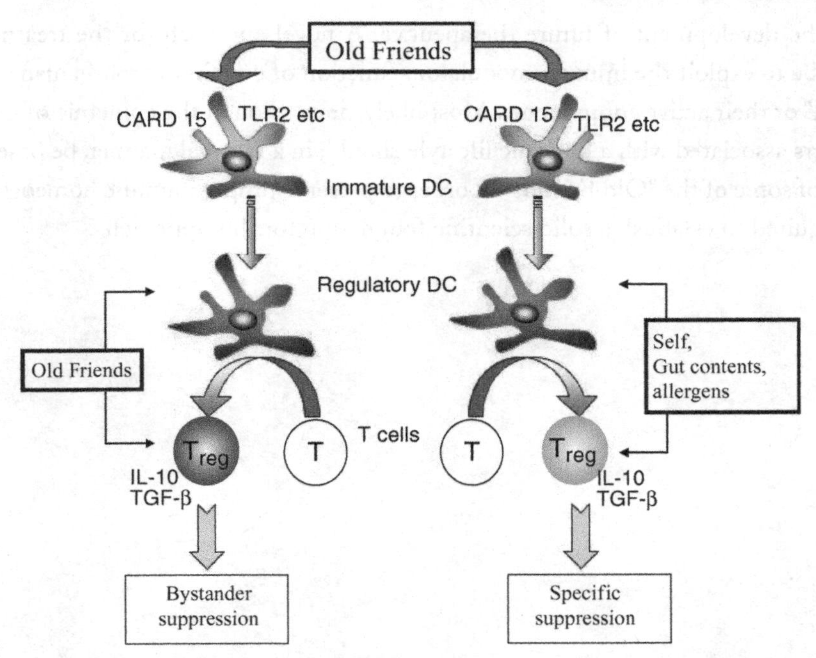

FIGURE 21: Organisms recognized as harmless by the innate immune system ("Old Friends") through pattern recognition receptors, such as CARD15 and TLR2, cause dendritic cells (DC) to mature into regulatory DC that drive Treg polarization. Some of these Treg cells will recognize the "old friends" themselves, and thus provide a continuous background bystander regulation. Regulatory dendritic cells, however, might also process and present epitopes from self, allergens and gut contents (antigens, bacterial DNA motifs, microbial heat shock proteins) and so drive specific immune-regulation. Source: Figure 3 in Guarner F, Bourdet-Sicard R, Brandtzaeg P, Gill HS, McGuirk P, van Eden W, Versalovic J, Weinstock JV, Rook GA (2006). Mechanisms of Disease: the hygiene hypothesis revisited. *Nat Clin Pract Gastroenterol Hepatol* 3: 275–284. Reproduced with permission of Nature Publishing Group.

millennia. However, such "domestic" commensals are mostly depleted from the westernized environment. The current absence of our "Old Friends" is probably associated with defects in the development of immunoregulatory pathways and may, on a background of genetic susceptibility, explain the increased prevalence of immunological disorders in societies with high hygienic standards.

Autoimmune diseases and asthma are worldwide a major cause of human morbidity. Treatment of these disorders generally relies heavily on immunosuppressive drugs, which lack specificity of action and can have serious side effects. Approaches based on understanding immunopathogenesis and designing interventions that inhibit exaggerated immune responses have considerable potential in the development of future therapeutics. A novel approach for the treatment of such diseases may be to exploit the immunomodulatory function of certain microorganisms, such as our "Old Friends," or their active components. Most likely, prevention of the epidemic of immunoregulatory disorders associated with a hygienic lifestyle should, in a rational manner, be based on the reintroduction of some of the "Old Friends" in order to preserve proper immune homeostasis. Further research is required to establish a solid scientific foundation for this approach.

• • • •

CHAPTER 5

Dysfunction of the Enteric Microbiota

Several disease states or disorders have been associated with changes in the composition or function of the enteric microbiota (Table 4). For instance, acute diarrhea is usually caused by pathogens that proliferate and invade or produce toxins. Antibiotic-associated diarrhea is due to imbalance in the composition and structure of the gut microbiota with overgrowth of pathogenic species, such as toxigenic *Clostridium difficile* strains and may cause diarrhea and colitis of variable severity. Gut bacteria may play a role in the pathogenesis of the irritable bowel syndrome (Bolino & Bercik, 2010). Symptoms of abdominal pain, bloating, and flatulence may be related to excessive production of gas by fermentations taking place in the distal small bowel and colon. Likewise, putrefaction of proteins by bacteria within the gut lumen is associated with the pathogenesis of hepatic encephalopathy in patients with acute or chronic liver failure.

Table 4 also includes autoimmunity and related disorders as potential consequences of dysfunction of gut microbial colonization. Gut bacteria play an important role in the instruction, development and homeostasis of the immune system. There is growing interest in understanding how microbial colonization may influence the onset and progression of autoimmunity and other immunoregulatory disorders such as atopies and inflammatory bowel diseases. Studies using molecular techniques confirm previous reports that the gut microbial ecosystem differs between children with and without atopic eczema. A large cohort study in the Netherlands suggests that differences in gut microbiota composition precede the development of atopic eczema. Early colonization by *Escherichia coli* increases risk of developing eczema and colonization with *Clostridium difficile* was associated with a higher risk of eczema, recurrent wheeze and allergic sensitization (Van Nimwegen et al., 2011).

Intestinal bacteria appear to play a role in the initiation of colon cancer through production of carcinogens, cocarcinogens or procarcinogens. The current evidence is mostly based in experimental models of colon carcinogenesis (Rowland, 2009). The molecular genetics of human colorectal cancer are well established, but epidemiological evidence suggests that environmental factors such as diet may play a major role in the development of sporadic colon cancer. Dietary fat and consumption of excess red meat, particularly processed meats, are associated with an enhanced risk in case-control studies. In contrast, elevated intake of fruits and vegetables, whole grain cereals, fish and calcium has been associated with reduced cancer risk. Dietary factors and genetic factors interact in part via

TABLE 4: Microbiota dysfunction and potential impacts on disease.

DISORDER	CLAIMED MICROBIOTA DYSFUNCTION
Infectious diarrhea, antibiotic-associated diarrhea	Altered composition/structure of microbial community
Septic complications: multisystem organ failure, diverticulitis, appendicitis	Deficient barrier function
Necrotizing enterocolitis	Altered composition/structure of microbial community Deficient barrier function
Hepatic encephalopathy	Metabolic dysfunction Deficient barrier function
Functional disorders: constipation, bloating, irritable bowel syndrome	Metabolic dysfunction Defects in trophic functions on motility, immunity
Obesity, type 2 diabetes, metabolic syndrome	Metabolic dysfunction Deficient barrier function Defects in trophic functions on immunity
Atopy, inflammatory bowel diseases, certain autoimmune disorders (?)	Defects in trophic functions on immunity
Colon cancer	Metabolic dysfunction: generation of genotoxic metabolites Defects in trophic functions on epithelial cells
Anxiety (?), autism (?)	Defects in trophic functions on central nervous system

events taking place in the lumen of the large bowel (Rafter et al., 2004). The influence of diet on the carcinogenic process may be mediated by changes in metabolic activities and the composition of the colonic microbiota. However, there is no information about abnormal composition of the enteric microbiota in relationship with colonic carcinogenesis.

The term "dysbiosis" refers to a condition characterized by imbalanced or abnormal microbial colonization, which may be associated with a health disorder. Even if the notion of a "normal" or "balanced" enteric microbiota is ambiguous and still undefined, studies are being conducted in order to identify microbial signatures associated with such disease states or disorders. Currently, most data refer to inflammatory bowel diseases and obesity, and they will be described in special sections of this chapter. In addition, another section will review the role of gut microbes in sepsis by bacterial translocation.

5.1 TRANSLOCATION

Gut bacteria are traditionally considered as a major source of infection and disease. Since the late 19th century, it is widely accepted that peritonitis may result from the passage of bacteria from the gut to the peritoneal cavity under certain conditions, as it is evident in cases of appendicitis, diverticulitis or gastrointestinal perforation. The passage of viable bacteria from the gastrointestinal tract through the epithelial mucosa is called bacterial translocation (Lichtman, 2001). Translocation of viable or dead bacteria in minute amounts constitutes an important boost to the mononuclear phagocyte system, especially to the Kupffer cells in the liver, and can be considered as a physiological mechanism for stimulation and instruction of the immune system. However, dysfunction of the gut mucosal barrier may result in translocation of a conspicuous quantity of viable microorganisms, usually belonging to Gram-negative aerobic genera (*Escherichia, Proteus, Klebsiella*). After crossing the epithelial barrier, bacteria may travel via the lymph to extraintestinal sites, such as the mesenteric lymph nodes, liver and spleen. Subsequently, enteric bacteria may disseminate throughout the body producing sepsis, shock, multisystem organ failure or death of the host. Extensive work on bacterial translocation has been performed in animal models and occurs notably in hemorrhagic shock, burn injury, trauma, intestinal ischemia, intestinal obstruction, severe pancreatitis, acute liver failure and cirrhosis. The three primary mechanisms promoting bacterial translocation in animal models are identified as follows: (a) small-bowel bacterial overgrowth, (b) increased permeability of the intestinal mucosal barrier and (c) deficiencies in host immune defenses (Lichtman, 2001).

Convincing evidence exists that bacterial translocation can occur in humans during various disease processes. Indigenous gastrointestinal bacteria have been cultured directly from the mesenteric lymph nodes of patients undergoing laparotomy. Data suggest that there may be a baseline rate of positive mesenteric lymph node culture approaching 5% in otherwise healthy humans. However, in conditions such as multisystem organ failure, acute severe pancreatitis, advanced liver cirrhosis, intestinal obstruction and inflammatory bowel diseases, observed rates of positive culture are much higher (16% to 40%). Bacterial translocation is associated with a significant increase in the development of postoperative sepsis in surgical patients (MacFie, 2004). Intestinal bacteria are involved in the development of multisystem organ failure in humans (Gatt et al., 2007). Massive release of proinflammatory mediators in response to translocating bacteria due to intestinal hypoperfusion is perceived as the key event (Figure 22). The gut becomes a proinflammatory organ, releasing chemokines, cytokines

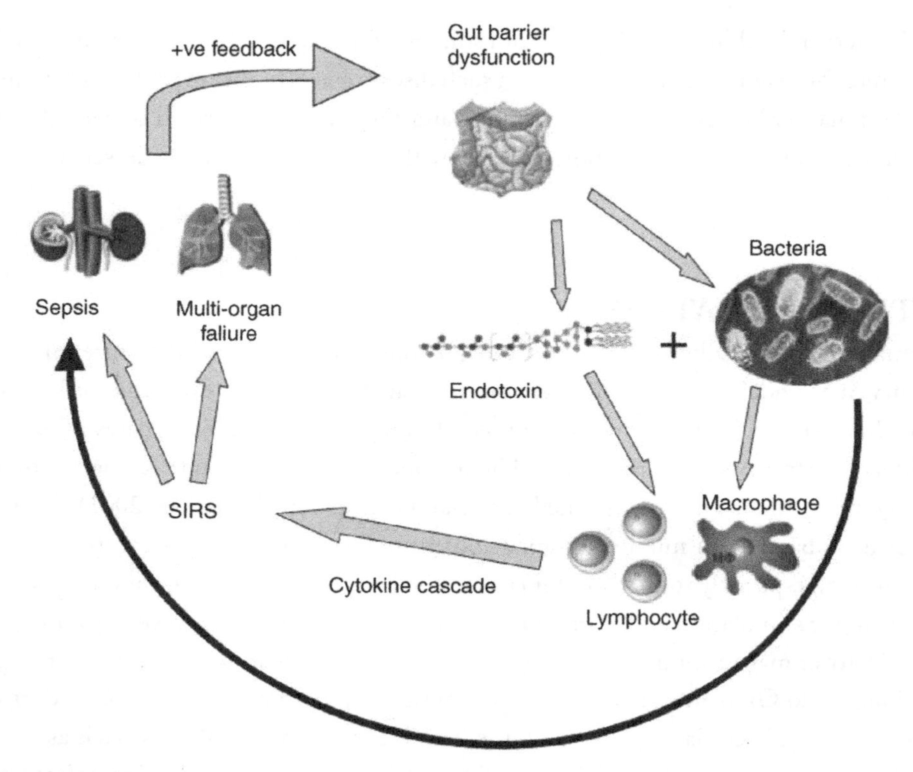

FIGURE 22: The gut origin of sepsis hypothesis, with bacterial translocation as a potential stimulus for ongoing inflammation. Source: Figure 1 in Gatt M, Reddy BS, MacFie J. Review article: bacterial translocation in the critically ill—evidence and methods of prevention. *Aliment Pharmacol Ther* 2007 Apr 1; 25(7): 741–57. Reproduced with permission of Wiley-Blackwell.

and other proinflammatory intermediates which affect both the local as well as the systemic immune systems, finally resulting in systemic inflammatory response syndrome (SIRS), and multisystem organ failure. Necrotizing enterocolitis in premature infants results from local bacterial translocation through the intestinal mucosa, or toxin-related injury of intestinal epithelia (Sherman, 2010). In cirrhotic patients, bacterial translocation may cause spontaneous bacterial peritonitis, an important complication of advanced liver disease. In this setting, the intestinal mucosa remains apparently intact but overgrowth of bacteria within the small bowel plays a major role (Guarner & Soriano, 2005).

Barrier dysfunction can be due to abnormal colonization of the gastrointestinal tract. Dysbiosis of the intestine, or abnormal gut microbiota, increases the risk necrotizing enterocolitis (Sherman, 2010). Infants developing necrotizing enterocolitis had less bacterial diversity, increased abundance of γ-*Proteobacteria* (e.g., *Enterobacteriacea*) and a decrease of *Firmicutes* in the stools

(Mai et al., 2011). Reasons for dysbiosis include the following: birth by cesarean section, hygiene practices, prolonged antibiotic administration, reduced bowel motility, immature epithelial host defenses and type or mode of nutrition.

5.2 INTESTINAL INFLAMMATION

There is convincing evidence that intestinal inflammation results from the interaction of the gut microbiota with the mucosal immune compartments. As described previously, the gastrointestinal tract is adapted to the analytical recognition of the external environment. The large mucosal surface is a sensitive interface that includes tools and structures allowing a detailed scrutiny of foreign bodies transiting along the tract. From a functional point of view, gut-associated lymphoid tissues generate either immuno-inflammatory responses for rejection of potential pathogens or non-inflammatory responses for tolerance towards dietary antigens and commensal non-pathogenic microbes.

In pathological conditions, such as the inflammatory bowel diseases (IBD), abnormal communication between gut microbial communities and the mucosal immune system is being incriminated as the core defect leading to mucosal lesions (Strober et al., 2007). Studies have shown that the gut microbiota is as an essential factor in driving inflammation in IBD (Sartor, 2008; Guarner, 2008). In Crohn's disease, fecal stream diversion induces inflammatory remission and mucosal healing in the excluded intestinal segment, whereas infusion of intestinal contents reactivates the disease. In ulcerative colitis, short-term treatment with an enteric-coated preparation of broad-spectrum antibiotics rapidly reduced metabolic activity of the microbiota and mucosal inflammation. These observations indicate that luminal bacteria provide the stimulus for immuno-inflammatory responses leading to mucosal injury (Sartor, 2008). In addition, patients show abnormal mucosal secretion of IgG antibodies against commensal bacteria (physiological response is based on IgA antibodies that do not trigger inflammation) (Macpherson, 1996), and mucosal T-lymphocytes are hyper-reactive against antigens of the common intestinal microbiota. Thus, in these patients local tolerance mechanisms towards commensal microbes seem to be abrogated (Guarner, 2008).

Despite the evidence that the microbiota is necessary to drive the inflammation, certain microbial species may indeed protect the mucosa from inappropriate host-damaging inflammatory responses. Studies in animal models and in organ culture of inflamed mucosal samples from patients with IBD have shown that some specific bacteria can down-regulate the expression of a number of key pro-inflammatory cytokines (TNFalpha, IFNgamma, IL-6, IL-23p19, IL-12p35, IL-17F) and chemokines (IL-8, CXCL1, CXCL2) (Borruel et al., 2002; Carol et al., 2006; Foligne et al., 2007; Llopis et al., 2009), thereby reducing the mucosal lesion scores (Llopis et al., 2005; Foligne et al., 2007). Moreover, some species of various genera (*Lactobacillus, Bifidobacterium* and *Faecalibacterium*) are able to stimulate immuno-regulatory IL-10 production (Borruel et al., 2003; Hart et al., 2004;

Sokol et al., 2008). Thus, some members of the gut microbial community may exacerbate inflammation but others can induce immuno-regulatory pathways that mitigate inflammatory reactions.

Altogether, several factors have been suggested to contribute to the loss of tolerance towards members of the indigenous microbiota in IBD patients, including genetic susceptibility, defects in mucosal barrier function and imbalance in the composition of the gut microbiota (excess of aggressive versus "friendly" commensal bacteria). Experts proposed that either primary dysregulation of the mucosal immune system leads to excessive immunologic responses to normal microbiota or changes in the composition of gut microbiota elicit pathologic responses from a normal mucosal immune system (Strober et al., 2007). The second alternative raised the hypothesis that an altered composition of the gut microbiota plays a key role in the pathogenesis of IBD, and it is currently being the focus of intensive research.

Infectious diseases are produced by specific microbial agents that possess the capacity of transmitting the disease to susceptible individuals. An infectious origin of inflammatory bowel diseases has never been confirmed, despite intensive search on a number of "candidate microorganisms" (*Mycobacterium*, *Listeria*, *Helicobacter* and *Mycoplasma* species, some *E. coli* strains, measles virus, etc.) (Marteau et al., 2004). Moreover, the lack of documented transmission among patients with Crohn's disease or ulcerative colitis makes it very unlikely that a single pathogen would be linked with the causation of any of these diseases.

However, it has been repeatedly demonstrated that the fecal microbiota differs between subjects with IBD and healthy controls (Guarner, 2008). Molecular techniques indicate that a substantial proportion of fecal bacteria (from 30% to 40% of dominant species) in patients with active Crohn's disease or ulcerative colitis belong to phylogenetic groups that are unusual in healthy subjects (Seksik et al., 2003). These remarkable changes could be secondary to disease activity but they are not observed in patients with infectious diarrhea. On the other hand, studies have shown reduced diversity of bacteria species in both fecal- and mucosa-associated communities in patients with IBD (Manichanh et al., 2006; Frank et al., 2007). These two studies employed 16S rRNA sequencing for exhaustive investigation of bacterial diversity in Crohn's disease and found a striking reduction of Firmicutes in patients as compared to healthy controls (Figure 23). A study in individuals with quiescent ulcerative colitis followed during one year observed that the composition of the microbiota was highly variable over time (Martinez et al., 2008). Temporal instability in the microbiota may be a consequence of low biodiversity and suggests that the intestinal ecosystem may be more susceptible to environmental influence. Temporal instability of dominant species has also been reported in Crohn's disease. Studies on mucosa-associated bacteria in IBD found higher concentrations of adherent bacteria in patients with clinically active disease, either ulcerative colitis or Crohn's disease, than in healthy controls (Swidsinski et al., 2002). This finding has also been reported in pediatric patients.

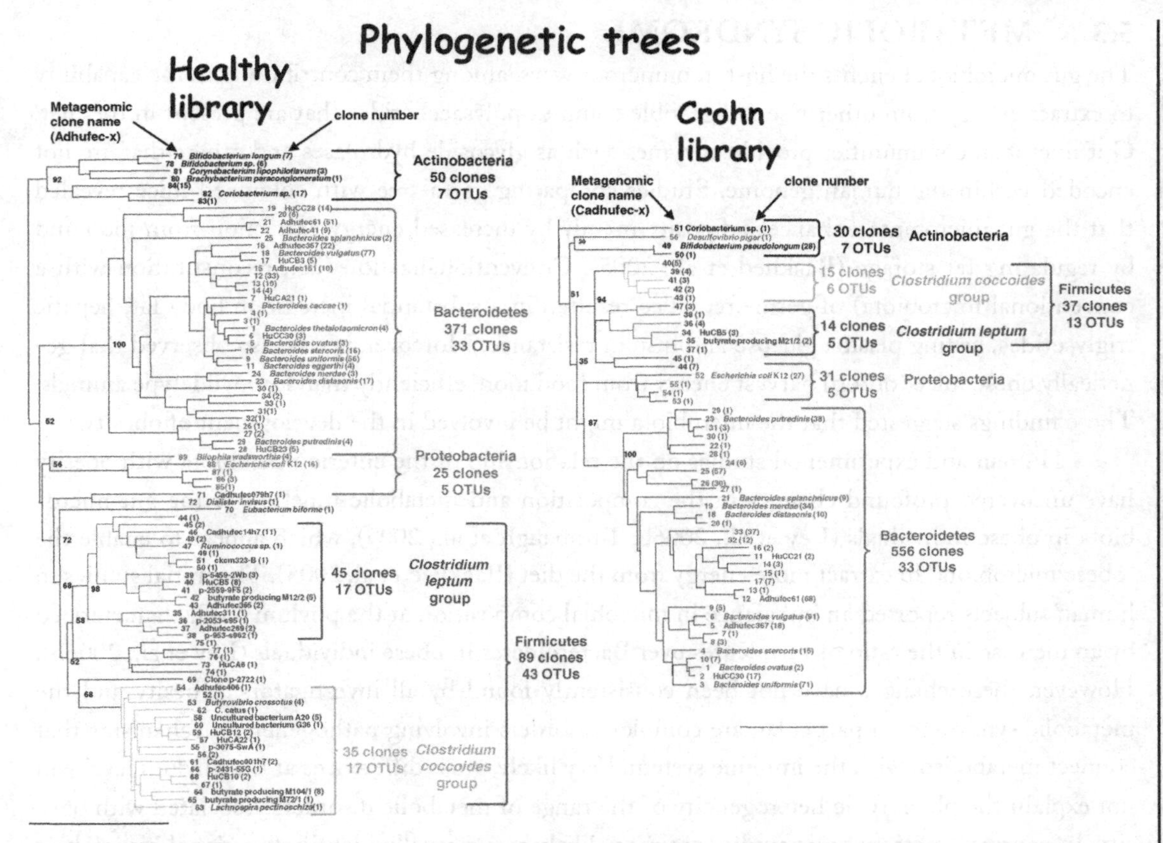

FIGURE 23: Comparison of the phylogenetic libraries derived from healthy subjects and patients with Crohn's disease in remission. The number of clones and phylotypes (OTUs) belonging to the Firmicutes phylum is strikingly reduced in Crohn's disease ($p < 0.025$). Source: combined from Figures 2 and 3 in Manichanh C, Rigottier-Gois L, Bonnaud E, Gloux K, Pelletier E, Frangeul L, Nalin R, Jarrin C, Chardon P, Marteau P, Roca J, Dore J. Reduced diversity of faecal microbiota in crohn's disease revealed by a metagenomic approach. *Gut* 2006; 55: 205–211. Reproduced with permission of BMJ Publishing Group Ltd.

In summary, enteric bacteria provide the stimulus for immuno-inflammatory responses leading to mucosal injury in IBD. There is evidence showing that the microbiota of patients with IBD differs from that of healthy subjects. Differences include low biodiversity of dominant bacteria, temporal instability and changes both in composition and spatial distribution: high numbers of adherent bacteria in the mucus layer and at the epithelial surface. Current data are compatible with the hypothesis that dysbiosis may be in the origin of the dysregulated immune response against some gut commensal microbes.

5.3 METABOLIC SYNDROME

The gut microbiota benefits the host in numerous ways, among them contributing in the capability to extract energy from otherwise indigestible complex polysaccharides that are present in the diet. Gut microbial communities provide enzymes such as glycoside hydrolases and others that are not encoded within the human genome. Studies comparing germ-free with colonized mice revealed that the gut microbiota enhances adiposity mainly by increased energy extraction from food and by regulating fat storage (Backhed et al., 2005). Conventionalization (i.e., reconstitution with a conventional microbiota) of germ-free mice resulted in a substantial increase in body fat, hepatic triglycerides, fasting plasma glucose and insulin resistance. Moreover, it was also observed that genetically obese mice (ob/ob) harvest energy from food more efficiently than lean wild-type animals. These findings suggested that the microbiota might be involved in the development of obesity.

Human and experimental studies on the relationship of the enteric microbiota with obesity have uncovered profound changes in the composition and metabolic function of the gut microbiota in obese individuals (Ley et al., 2006b; Turnbaugh et al., 2009), which appear to enable the "obese microbiota" to extract more energy from the diet (Backhed et al, 2005). The initial studies in human subjects reported an imbalance in microbial composition at the phylum level, characterized by an increase in the ratio of Firmicutes over Bacteroidetes in obese individuals (Ley et al., 2006b). However, these changes have not been consistently found by all investigators. Obesity and the metabolic syndrome, in particular, are complex disorders involving pathogenetic mechanisms that connect metabolism with the immune system. Very likely, gross differences at the phylum level will not explain the phenotypic heterogeneity of the range of metabolic disorders associated with obesity. In contrast, metagenomic studies are more likely to eventually identify functional biomarkers correlating strongly with phenotype (Tilg & Kaser, 2011).

Obesity and insulin resistance are associated with low-grade systemic inflammation (Cani & Delzenne, 2009). Gut microbial communities seem to play a role in the origin of the inflammatory disturbance associated with the metabolic syndrome. In fact, endotoxin lipopolysaccharides (LPS) of gram-negative bacteria have been considered to be the triggering factor for the early development of low-grade inflammation leading to insulin resistance (Cani et al., 2007). Excessive intake of dietary fat would facilitate the absorption of pro-inflammatory bacterial LPS from the gut, thereby increasing plasma LPS level leading to subclinical metabolic endotoxemia. This hypothesis has recently been tested in an animal model (Figure 24). Interestingly, when metabolic endotoxemia was reproduced in mice through continuous subcutaneous infusion of LPS for 4 weeks, fasted glycemia and insulinemia, as well as adipose tissue and whole body weight were increased to a similar extent as in high-fat-fed mice (Cani et al., 2007). In addition, parallel experiments showed that obesity induced by diet (high-fat feeding) or by genetic deletion (leptin-deficient models) (Waldram et al., 2009) is associated with the changes in gut microbiota characterized by a decreased number of bifi-

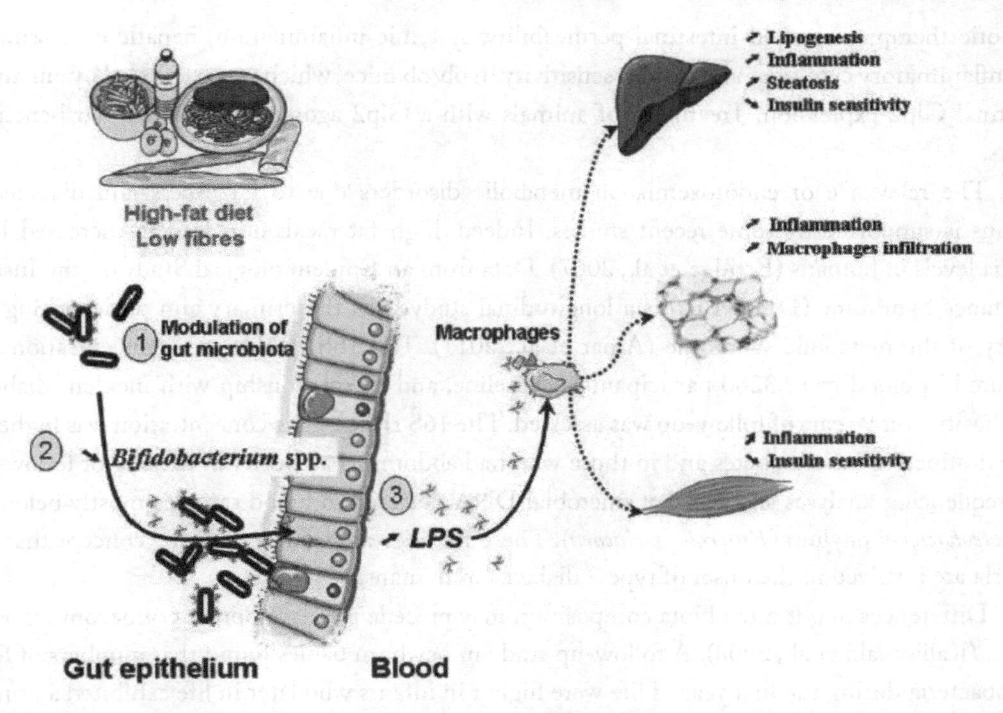

FIGURE 24: High-fat diet feeding changes gut microbiota, promotes metabolic endotoxemia and triggers the development of metabolic disorders. High-fat diet feeding of mice decreases bifidobacteria numbers, and increases plasma LPS levels and secretion of proinflammatory cytokines. High fat feeding promotes low-grade inflammation-induced metabolic disorders (insulin resistance, diabetes, obesity, liver steatosis). Source: Figure 4 in Cani PD & Delzenne NM. The role of the gut microbiota in energy metabolism and metabolic disease. *Current Pharmaceutical Design*, 2009, 15, 1546–1558. Reproduced with permission of Bentham Open.

dobacteria. Subsequent experiments tested the effect of feeding mice with inulin-type fructans. The prebiotic restored the number of intestinal bifidobacteria and reduced the impact of high-fat diet on metabolic endotoxemia and low-grade inflammatory disturbance (Cani et al., 2009). Prebiotic feeding also improved glucose tolerance and glucose-induced insulin secretion in high-fat diet mice.

High-fat diets may affect epithelial integrity and hence lead to impaired gut permeability, and consequently to systemic inflammation via translocation. Prebiotic carbohydrates or antibiotics lowered systemic endotoxin levels and inflammatory cytokine expression in the liver (Cani et al., 2008). Such improvement of metabolic inflammation in obese mice might not only involve changes in the microbiota but also expression of glucagon-like peptide 2 (Glp2), an intestinal growth factor with anti-inflammatory activities that stabilizes intestinal barrier function (Cani et al., 2009).

Prebiotic therapy improved intestinal permeability, systemic inflammation, hepatic expression of pro-inflammatory cytokines and insulin sensitivity in ob/ob mice, which was paralleled by enhanced intestinal Glp2 expression. Treatment of animals with a Glp2 agonist revealed similar beneficial effects.

The relevance of endotoxemia on metabolic disorders due to fat excess and diabetes in humans is supported by some recent studies. Indeed, high-fat meals also lead to increased LPS plasma levels in humans (Erridge et al., 2007). Data from an Epidemiological Study on the Insulin Resistance Syndrome (D.E.S.I.R.) is a longitudinal study with the primary aim of describing the history of the metabolic syndrome (Amar et al., 2011). The 16S rRNA gene concentration was measured in blood from 3280 participants at baseline, and its relationship with incident diabetes and obesity over 9 years of follow-up was assessed. The 16S rRNA gene concentration was higher in those destined to have diabetes and in those who had abdominal adiposity at the end of follow-up. Pyrosequencing analyses showed that microbial DNA detected in blood samples mostly belonged to *Proteobacteria* phylum (*Enterobacteriacaea*). These findings are evidence for the concept that gut bacteria are involved in the onset of type 2 diabetes in humans.

Differences in gut microbiota composition may precede the development of becoming overweight (Kalliomaki et al., 2008). A follow-up study in newborn babies found that numbers of fecal bifidobacteria during the first year of life were higher in infants who later in life exhibited a normal weight than in infants becoming overweight. However, the impact of the prebiotic approach on endotoxemia and inflammation in obese and diabetic patients has not yet been demonstrated. Intervention studies with prebiotics are necessary to proof the role of gut microbial composition in the onset of insulin resistance and metabolic syndrome in humans.

• • • •

CHAPTER 6

Therapeutic Manipulation of the Enteric Microbiota

Symbiosis between enteric microbiota and host can be optimized by pharmacological or nutritional intervention on the gut microbial ecosystem using probiotics or prebiotics. Administration of exogenous bacteria with known properties may improve specific functions of the gut microbiota (metabolism, protection and trophism) or prevent dysfunction associated with disease. These bacteria are called "probiotics," a term that refers to "live micro-organisms which when administered in adequate amounts confer a health benefit on the host," as proposed by the Joint FAO/WHO Expert Consultation (2001). This definition was also adopted by the International Scientific Association for Probiotics and Prebiotics (Reid et al., 2003).

The term prebiotic refers to "a selectively fermented ingredient that allows specific changes, both in the composition and/or activity in the gastrointestinal microbiota that confers benefits upon host well being and health." A prebiotic should not be hydrolyzed by human intestinal enzymes, it should be selectively fermented by beneficial bacteria, and this selective fermentation should result in beneficial effects on health or well-being of the host (Gibson et al., 2004). The combination of probiotics and prebiotics is termed "synbiotic," and is an exciting concept aimed at optimizing the impact of probiotics on the gut microbial ecosystem.

Interventions on the gut microbial ecosystem also include the use of antibiotics and fecal transplant therapies. The use of antibiotics will be discussed in a special section. Fecal transplant, also called fecal bacteriotherapy, is emerging as a rather controversial alternative approach to manipulate the gut microbiota. A total of 239 patients who had undergone fecal transplant have been reported (Landy et al., 2011). Seventeen of 22 studies of fecal transplant were in fulminant or refractory *Clostridium difficile* infection. When standard treatment has failed, this approach seems to be an effective alternative therapy for patients with *Clostridium difficile* infection, showing disease resolution in 87% of cases. No adverse effects of fecal transplantation have been reported. Small numbers of patients are reported to have undergone successful fecal transplant for irritable bowel syndrome and inflammatory bowel disease. However, reports to date may suffer from reporting bias of positive outcomes and under-reporting of adverse effects.

6.1 ANTIBIOTICS

Antibiotics are useful drugs to combat infections and are widely used for treatment of infectious conditions due to bacterial translocation, including diverticulitis, spontaneous bacterial peritonitis and sepsis. Non-absorbable oral antibiotics are being used with success to revert hepatic encephalopathy. Other current indications are small-bowel bacterial overgrowth and some specific gastrointestinal infections. Recently, the use of non-absorbable oral antibiotics has been proposed as treatment for alleviating symptoms of patients with irritable bowel syndrome without constipation (Pimentel et al., 2011).

It is important to mention that antibiotic overuse has led to an increase in bacterial resistance to treatments. Other serious long-term consequences of antibiotics misuse or overuse derive from the lack of selectivity in their effects. Antibiotics kill the targeted pathogen, as well as many other bystander commensal bacteria. Data suggest that some of these bacteria in the gut microbiota may never fully recover (Blaser, 2011). The growing incidence of *Clostridium difficile* infections is causally related to perturbations to the enteric microbiota by antibiotics. Long-term effects of antibiotics may even increase susceptibility to other infections and disease. Overuse of antibiotics could be fuelling the dramatic increase in conditions such as obesity, type 1 diabetes, inflammatory bowel disease, allergies and asthma (Blaser, 2011).

6.2 PROBIOTICS

Human and experimental studies with probiotics have targeted specific health benefits associated with the three functional areas of the gut microbiota (metabolic effects, protective effects and trophic effects), and worldwide research on this topic has accelerated in recent years. The scientific basis for the use of probiotic microbes was reviewed in depth by a task force of the American Academy of Microbiology (Walker & Buckley, 2006). In most human studies published so far, probiotics were administered alive either by oral route (as a food component or in the form of specific preparations of viable micro-organisms) or in topical preparations (skin, nasal or vaginal applications).

The use of probiotics in human nutrition and medicine is gradually growing. The Cochrane Central Register of Controlled Trials includes a large list of human studies that had tested probiotic efficacy and several systematic reviews. A major area for probiotic applications has been the prevention or treatment of gastroenterological diseases. There is ample evidence supporting the efficacy of some probiotic strains in several acute gastroenterological disorders (Figure 25). Many of these indications are already accepted in clinical practice. The World Gastroenterology Organization published a practical guideline that is available at the WGO website (www.worldgastroenterology.org). Recommendations by the guideline are summarized in this chapter section.

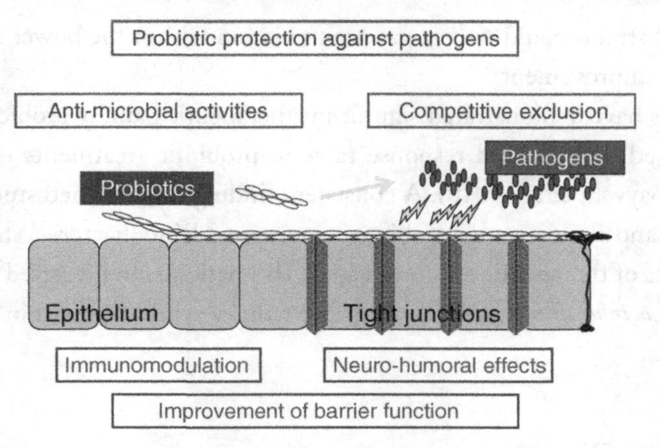

FIGURE 25: Probiotics can protect against gastrointestinal infections by different mechanisms of action. Some strains have effects at the intestinal luminal site and produce bacteriocins with antimicrobial activity against pathogens or compete with them for the available substrates and niches. Some strains have effects on the mucosal site by strengthening barrier function (mucus secretion, tight junction improvement) or stimulating immune defense. Source: author's drawing.

6.2.1 Lactose Digestion

Lactose malabsorption is the result of lactase deficiency in brush border epithelial cells of the small intestine. Due to this deficiency, a fraction of the ingested lactose is not absorbed in the small intestine. Subjects may develop gastrointestinal symptoms such as diarrhoea, flatulence, abdominal bloating and pain after ingestion of lactose containing food. Prevalence of lactase deficiency in adult populations is relatively high and varies between 5% to 15% in Northern European and American countries and 50% to 100% in African, Asian and South American countries. These subjects tend to eliminate milk and dairy products from their diet, and their calcium intake may be compromised. The bacteria used as starter culture in yogurt (*Streptococcus thermophilus* and *Lactobacillus delbrueckii* subsp. *bulgaricus*) can improve lactose digestion and eliminate symptoms in lactase-deficient individuals.

6.2.2 Irritable Bowel Syndrome

Colonic fermentations result in the generation of variable gas volumes in the intestine. However, some gut bacteria degrade metabolic substrates without producing gas, and some microbial species may consume gas, particularly hydrogen. Symptoms of abdominal pain, bloating and flatulence are commonly seen in patients with irritable bowel syndrome (IBS). Hypothetically, administration of appropriate bacterial strains could reduce gas accumulation within the bowel in these patients and induce symptomatic improvement.

Several studies have demonstrated significant therapeutic gain by probiotics when compared to placebo, as assessed by increased response rates to probiotic treatments or enhanced relief in symptom scores (Moayyedi et al., 2010). A consistent finding in published studies is a reduction of abdominal bloating and flatulence by probiotic treatments. Bifidobacterial strains appear to contribute to higher rates of therapeutic success in adult IBS patients. In breastfed babies with infantile colic, the probiotic *Lactobacillus reuteri* may improve colicky symptoms within 1 week of treatment (Savino et al., 2010).

6.2.3 Prevention of Diarrhea

Clinical trials have tested the efficacy of probiotics in the prevention of acute diarrheal conditions, including antibiotic-associated diarrhea, nosocomial- and community-acquired infectious enteritis and traveler's diarrhea (Sazawal et al., 2006). Different probiotics show promise as effective therapies for the prevention of antibiotic-associated diarrhea. Both the short- and long-term use of antibiotics can result in diarrhea, particularly during regimens with multiple drugs. In placebo-controlled studies, diarrhea occurred at a rate of 15% to 26% in the placebo arms but only in 3% to 7% of patients receiving a probiotic. Different strains have been tested including *Lactobacillus rhamnosus* strain GG, *Lactobacillus acidophilus*, *Lactobacillus casei*, *Bacillus clausii*, the yeast *Saccharomyces boulardii* and others.

6.2.4 Treatment of Acute Diarrhea

Probiotics are useful as treatment of acute infectious diarrhea in children. Different strains, including *L. reuteri, L. rhamnosus GG, L. casei* and *S. boulardii*, have been tested in controlled clinical trials and were proven useful in reducing the severity and duration of diarrhea. Oral administration of probiotics shortens the duration of acute diarrheal illness in children by approximately 1 day. The results of the systematic reviews are consistent and suggest that probiotics are safe and effective (Allen et al., 2010; Szajewska et al., 2007).

6.2.5 Eradication of *Helicobacter pylori*

Probiotics have been tested as a strategy for eradication of *Helicobacter pylori* infection of the gastric mucosa in humans. Clinical studies have tested the efficacies of different probiotic strains in combination with standard therapies with antibiotics. Adding probiotics can increase eradication rates after triple or quadruple anti-*H. pylori* antibiotic regimens (Tong et al., 2007).

6.2.6 Prevention of Systemic Infections

Necrotizing enterocolitis (NEC) is a severe clinical condition that may occur in low birth weight neonates due to relative immaturity and dysfunction of the gut mucosal barrier. Several controlled studies have demonstrated that the use of probiotic mixtures in low birth weight infants significantly reduces the incidence and severity of necrotizing enterocolitis and may prevent mortality. A meta-analysis of the published trials suggests that probiotics reduced the risk of developing necrotizing enterocolitis by two thirds and risk of death by one half (Deshpande et al., 2010).

6.2.7 Inflammatory Bowel Diseases

In ulcerative colitis, three randomized controlled trials investigated the effectiveness of an orally administered enteric-coated preparation of viable *Escherichia coli* strain Nissle 1917 as compared with mesalazine, the standard treatment for maintenance of remission (Kruise et al., 2004). These studies concluded that this strain has an equivalent effect to mesalazine in maintaining remission.

The VSL#3 mixture has been proven highly effective for maintenance of remission of chronic relapsing pouchitis, after induction of remission with antibiotics (Gionchetti et al., 2000). This probiotic mixture is also used for treatment of mildly active ulcerative colitis (Tursi et al., 2010).

6.3 PREBIOTICS

Some microbial species living in the gut are potential pathogens and may cause disease when the integrity of the mucosal barrier is functionally breached. However, some other microbial species have never been incriminated in human disease (Borriello et al., 2003). Since the pioneering conception by Nobel Laureate Elie Metchnikoff in the early 20th century (Metchnikoff, 1907), experts have repeatedly suggested that "a healthy or balanced gut microbiota is one that is predominantly saccharolytic and comprises significant numbers of bifidobacteria and lactobacilli" (Cummings et al., 2004). Taking into account that food ingested by the host is a major nutritional source for the microbial communities living in the gut, it is therefore plausible that unabsorbed portions of food may influence activities and composition of the microbial communities in the gut.

Early research in the 1980s and 1990s demonstrated that specific non-digestible oligosaccharides were selectively fermented by bifidobacteria and had the capacity of increasing bifidobacteria counts in human faeces (Gibson et al., 1995). These observations led to the introduction of the concept of prebiotics by Gibson & Roberfroid (1995). Prebiotics were initially defined as "non-digestible food ingredients that beneficially affect the host by selectively stimulating the growth and/or activity of one or a limited number of bacteria in the colon, and thus improves host health." Complex polysaccharides, such as dietary fiber, are not hydrolyzed by human digestive enzymes but should not necessarily be ascribed to the prebiotic concept as they may not have selective effects on the gut microbiota. For this and other reasons, the authors updated the prebiotc concept some years later and proposed three criteria for classifying a food ingredient as a prebiotic. Accordingly, a prebiotic (1) should not be hydrolyzed by human intestinal enzymes, (2) it should be selectively fermented by beneficial bacteria and (3) this selective fermentation should result in beneficial effects on health or well-being of the host (Gibson et al., 2004). The authors refined the definition and proposed that a prebiotic is "a selectively fermented ingredient that allows specific changes, both in the composition and/or activity in the gastrointestinal microbiota that confers benefits upon host well being and health" (Gibson et al., 2004). The prebiotic concept has attracted the interest of many academic as well as industrial scientists, and it has become a popular research topic in nutrition and, more recently, in the biomedical fields.

The majority of the scientific data on prebiotic effects have been obtained using food ingredients belonging to two chemical groups namely inulin-type fructans and the galacto-oligosaccharides (see Table 5). Inulin-type fructans are the most extensively studied prebiotics, and consist of polymers of fructose linked by beta 2-1 fructosyl-fructose bonds. A starting glucose moiety can be present but it is not necessary. The galacto-oligosaccharides consist of a lactose core with one or more galactosyl residues linked via beta 1-3, beta 1-4 and beta 1-6 bonds. These products have repeatedly demonstrated the capacity to selectively stimulate the growth of bifidobacteria and, in some cases, lactobacilli leading to a significant change in gut microbiota composition (Roberfroid et al., 2010).

Inulin-type fructans are naturally present as energy reserve in a wide variety of plants. These are cereals such as wheat and barley, vegetables like onion and garlic, and fruits such as banana and tomato. Consumption of such edible vegetal products is a natural source of prebiotics in Western-style diets (Van Loo et al., 1995). The daily per capita intake is estimated to range from 1 to 10 g of inulin-type fructans, depending on geographic, demographic, and other related parameters (age, sex, season, etc.). Inulin-type fructans content in foods is not measurable by classic methods of dietary fiber analysis and consequently this information may not be mentioned in food tables. Inulin is also present in plants that are not commonly used in human diets, for instance in chicory plants (Belgian endives, Figure 26), particularly in their roots (16% to 18% by weight).

Commercially available galacto-oligosaccharides are produced from lactose, but some of them are naturally present in the human milk. The characteristic composition of human milk is associated with a bifidogenic effect in breast fed neonates (Coppa et al., 2004). The human milk is rich in oligosaccharides that are resistant to digestive processes and reach the colon. Cow milk and human milk have significant qualitative and quantitative differences regarding these carbohydrates, and the bifidogenic effect of human milk is not observed in infants fed with cow's milk-based formulas. Because of their peculiar chemical structure, human milk oligosaccharides have a very significant role in modulating the intestinal microbiota of neonates.

TABLE 5: Description of products with established prebiotic effect.

INULIN-TYPE FRUCTANS LINEAR β (2→1) FRUCTOSYL-FRUCTOSE POLYMERS. GLUCOSE-FRUCTOSE$_n$ AND/OR FRUCTOSE-FRUCTOSE$_n$	
Short to large size polymers (DP 2–60)	Inulin (especially chicory inulin) (DP$_{av}$ 12)
Short oligomers (DP 2-8)	Fructo-oligosaccharides FOS scFOS (enzymatic synthesis from sucrose) (DP$_{av}$ 3–6) Oligofructose (enzymatic partial hydrolysis of inulin) (DP$_{av}$ 4)
Large-size polymers (DP 10–60)	High molecular weight inulin (physical purification) (DP$_{av}$ 25) lcFOS
Mixture (DP 2–8) + (DP 10–60)	Mixture of oligomers and large size polymers
GALACTANS MIXTURE OF β (1→6); β (1→3); β (1→4) GALACTOSYL-GALACTOSE	
Disaccharide Galactosyl-Fructose	Lactulose
Short oligomers Galactose$_n$-Galactose, and/ or Galactose$_n$-Glucose (DP2-8)	Galacto-oligosaccharides GOS Trans-galactooligosaccharides TOS (enzymatic transgalactosylsation of lactose)
Mixture of galactans and inulin-type fructans GOS-FOS	Galacto-oligosaccharides and high molecular weight inulin, Usually known as GOS-FOS or scGOS-lcFOS

DP, degree of polymerization; DP$_{av}$, average degree of polymerization; ITF, inulin-type fructans; FOS, fructo-oligosaccharides; scFOS, short-chain fructo-oligosaccharides; lcFOS, long-chain fructo-oligosaccharides; GOS, galacto-oligosaccharides; scGOS, short-chain galacto-oligosaccharides; TOS, trans-galacto-oligosaccharides. Adapted from Roberfroid et al. (2010).

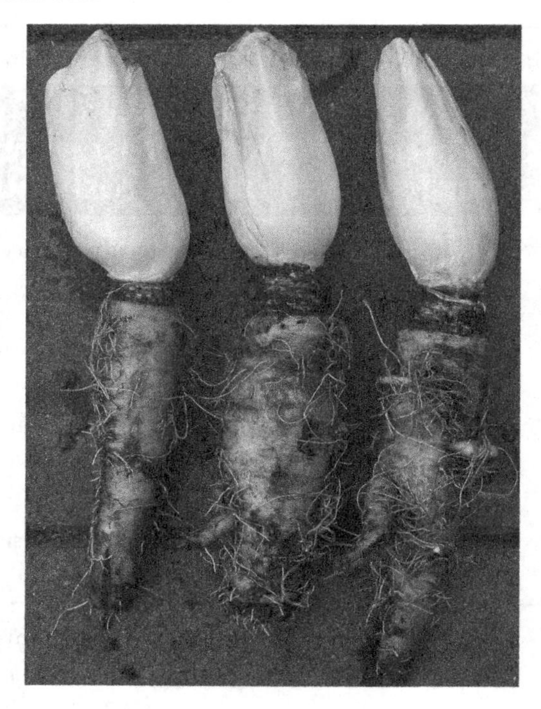

FIGURE 26: The root of Belgian endives contains up to 20% inulin, a non-digestible polysaccharide with prebiotic effects. Source: Wikipedia.

Human trials have repeatedly confirmed the prebiotic effect of inulin-type fructans and galacto-oligosaccharides as evidenced by their ability to change the gut microbiota composition after a short feeding period. At doses starting from around 5 grams per day, these compounds increase the number of fecal bifidobacteria by one or two logarithm orders in most individuals (Roberforid et al., 2010). It has been observed that he daily dose of a prebiotic does not correlate with the absolute numbers of "new" bacterial cells that appear as a consequence of the prebiotic treatment. The initial numbers of bifidobacteria seems to influence the prebiotic effect and those subjects with initial low counts are the best responders after prebiotic feeding (De Preter et al., 2008). Interestingly, changes have also been shown in the mucosa-associated microbiota (Langlands et al., 2004).

Oligofructose-enriched inulin and lactulose have been shown to reduce microbial fermentation of proteins in the human colon (De Preter et al., 2007; De Preter et al., 2008), as consequence of enhanced saccharolytic activity. Inulin-type fructans have positive effects on basic physiological functions of the colon, i.e., stool production and fecal excretion (Roberfroid, 2005), and are being used for bowel regularity.

Prebiotics may influence the immune system directly or indirectly either via specific products derived from carbohydrate fermentation or by altering microbial composition of the gut microbiota. Increased numbers of a particular microbial genus or species, or relative decrease of other microbial species, may change the collective interaction of the microbiota on immune sensors and shift the balance between effector and regulatory pathways (Guarner et al., 2006). On the other hand, microbial products such as SCFAs, defensins or other bioactive peptides may interact with immunocompetent cells, including enterocytes, and modify their status of activation and/or their activity. For instance, some G-protein coupled receptors for SCFA are expressed on leukocytes, especially in polymorphonuclear cells (Le Poul et al., 2003), as well as on epithelial cells of the human colon (Karaki et al., 2008). SCFAs modulate chemokine expression in intestinal epithelial cells (Sanderson, 2007), and are critical for controlling inflammatory responses and maintaining intestinal homeostasis (Maslowski et al., 2009). Finally, some oligosaccharides including oligo-fructose can bind to cell receptors on pathogenic bacteria and avoid them to attach to epithelial cells preventing invasion.

Studies in healthy infants have shown that inulin-type fructans and galacto-saccharides increase the luminal secretion of total secretory IgA into the intestinal lumen (Scholtens et al., 2008). In mice, the number of B lymphocytes in Peyer's patches was shown to increase in parallel to IgA changes. Those studies also showed enhanced IL-10 cytokine secretion by intestinal tissue and decreased transcription and concentration of pro-inflammatory cytokines (Roller et al., 2004). There may be a greater effect in young individuals as their gut immune system is still developing and may therefore be more susceptible to modulation (Lomax & Calder, 2009).

Prebiotics have been tested in humans for their ability to prevent infectious diseases, as this outcome would be a relevant consequence of improving immune function. Episodes of common childhood diarrhea were reported to be reduced in healthy infants supplemented with oligo-fructose (Waligora-Dupriet et al., 2007). Incidence of acute diarrhea was also reduced in infants who received an infant formula containing GOS/FOS (Bruzzese et al., 2009). In adults, oligo-fructose supplements were reported useful for decreasing relapse rate of *Clostridium difficile*-associated diarrhoea (Lewis et al., 2005).

Specific indications of prebiotics for human health include their use in infant formulas. The effect of breast feeding on infant gut microbiota composition is well established, and mother's milk is known to contain complex mixtures of oligosaccharides with prebiotic bifidogenic effects. Therefore, infant formulas and foods have been supplemented with prebiotics, and such supplementation increases the faecal concentration of bifidobacteria.

Dietary intake of inulin-type fructans or lactulose has been shown to increase calcium absorption as well as bone calcium accretion improving bone mineral density, especially in adolescents (Roberfroid et al., 2010).

Finally, interesting data, mainly from experimental models, support the beneficial effects of inulin-type fructans on energy homeostasis, satiety regulation and body weight gain (see Chapter 5). These studies suggest that the gut microbiota composition (especially the number of bifidobacteria) may contribute to modulate metabolic processes associated with the metabolic syndrome. A leading role of altered gut permeability and metabolic endotoxemia in the pathogenesis insulin resistance and diabetes has been reported. Prebiotic food ingredients have been shown to reverse these derangements in the animal model.

• • • •

References

Abrams, G. D., Bishop, J. E. Effect of the normal microbial flora on the resistance of the small intestine to infection. *J Bacteriol* **92**(6): pp. 1604–8, 1966 Dec.

Allen, S. J., Martinez, E. G., Gregorio, G. V., Dans, L. F. Probiotics for treating acute infectious diarrhoea. Cochrane Database of Systematic Reviews, Issue 11. Art. No.: CD003048, 2010.

Amar, J., Serino, M., Lange, C., Chabo, C., Iacovoni, J., Mondot, S., Lepage, P., Klopp, C., Mariette, J., Bouchez, O., Perez, L., Courtney, M., Marre, M., Klopp, P., Lantieri, O., Doré, J., Charles, M. A., Balkau, B., Burcelin, R., for the D.E.S.I.R. Study Group. Involvement of tissue bacteria in the onset of diabetes in humans: evidence for a concept. *Diabetologia* [Epub ahead of print] PubMed PMID: 21976140, 2011 Oct 6.

Arumugam, M., Raes, J., Pelletier, E., Le Paslier, D., Yamada, T., Mende, D. R., Fernandes, G. R., Tap, J., Bruls, T., Batto, J. M., Bertalan, M., Borruel, N., Casellas, F., Fernandez, L., Gautier, L., Hansen, T., Hattori, M., Hayashi, T., Kleerebezem, M., Kurokawa, K., Leclerc, M., Levenez, F., Manichanh, C., Nielsen, H. B., Nielsen, T., Pons, N., Poulain, J., Qin, J., Sicheritz-Ponten, T., Tims, S., Torrents, D., Ugarte, E., Zoetendal, E. G., Wang, J., Guarner, F., Pedersen, O., de Vos, W. M., Brunak, S., Doré, J., MetaHIT Consortium, Antolín, M., Artiguenave, F., Blottiere, H. M., Almeida, M., Brechot, C., Cara, C., Chervaux, C., Cultrone, A., Delorme, C., Denariaz, G., Dervyn, R., Foerstner, K. U., Friss, C., van de Guchte, M., Guedon, E., Haimet, F., Huber, W., van Hylckama-Vlieg, J., Jamet, A., Juste, C., Kaci, G., Knol, J., Lakhdari, O., Layec, S., Le Roux, K., Maguin, E., Mérieux, A., Melo Minardi, R., M'rini, C., Muller, J., Oozeer, R., Parkhill, J., Renault, P., Rescigno, M., Sanchez, N., Sunagawa, S., Torrejon, A., Turner, K., Vandemeulebrouck, G., Varela, E., Winogradsky, Y., Zeller, G., Weissenbach, J., Ehrlich, S. D., Bork, P. Enterotypes of the human gut microbiome. *Nature* **473**(7346): pp. 174–80, 2011 May 12.

Atarashi, K., Tanoue, T., Shima, T., Imaoka, A., Kuwahara, T., Momose, Y., Cheng, G., Yamasaki, S., Saito, T., Ohba, Y., Taniguchi, T., Takeda, K., Hori, S., Ivanov, I. I., Umesaki, Y., Itoh, K., Honda, K. Induction of colonic regulatory T cells by indigenous Clostridium species. *Science* **331**(6015): pp. 337–41, 2011 Jan 21.

Bach, J. F. The effect of infections on susceptibility to autoimmune and allergic diseases. *N Engl J Med* **347**: pp. 911–20, 2002.

Bäckhed, F., Ley, R. E., Sonnenburg, J. L., Peterson, D. A., Gordon, J. I. Host-bacterial mutualism in the human intestine. *Science* **307**(5717): pp. 1915–20, 2005 Mar 25.

Bager, P., Wohlfahrt, J., Westergaard, T. Caesarean delivery and risk of atopy and allergic disease: meta-analyses. *Clin Exp Allergy* **38**: pp. 634–42, 2008.

Bager, P., Simonsen, J., Nielsen, N. M., Frisch, M. Cesarean section and offspring's risk of inflammatory bowel disease: a National Cohort Study. *Inflamm Bowel Dis* doi: 10.1002/ibd.21805. [Epub ahead of print] PMID: 21739532, 2011 Jul 7.

Bernstein, C. N., Shanahan, F. Disorders of a modern lifestyle: reconciling the epidemiology of inflammatory bowel diseases. *Gut* **57**(9): pp. 1185–91, 2008 Sept.

Blaser, M. Antibiotic overuse: Stop the killing of beneficial bacteria. *Nature* **476**(7361): pp. 393–4, 2011 Aug 24.

Bolino, C. M., Bercik, P. Pathogenic factors involved in the development of irritable bowel syndrome: focus on a microbial role. *Infect Dis Clin North Am* **24**(4): pp. 961–75, 2010 Dec.

Bouskra, D., Brézillon, C., Bérard, M., Werts, C., Varona, R., Boneca, I. G., Eberl, G. Lymphoid tissue genesis induced by commensals through NOD1 regulates intestinal homeostasis. *Nature* **456**: pp. 507–10, 2008.

Borriello, S. P., Hammes, W. P., Holzapfel, W., Marteau, P., Schrezenmeir, J., Vaara, M., Valtonen, V. Safety of probiotics that contain lactobacilli or bifidobacteria. *Clin Infect Dis* **36**: pp. 775–80, 2003.

Borruel, N., Carol, M., Casellas, F., Antolin, M., de Lara, F., et al. Increased mucosal tumour necrosis factor alpha production in Crohn's disease can be downregulated ex vivo by probiotic bacteria. *Gut* **51**: pp. 659–64, 2002.

Borruel, N., Casellas, F., Antolin, M., Llopis, M., Carol, M., et al. Effects of nonpathogenic bacteria on cytokine secretion by human intestinal mucosa. *Am J Gastroenterol* **98**: pp. 865–70, 2003.

Brandtzaeg, P. Mucosal immunity: induction, dissemination, and effector functions. *Scand J Immunol* **70**(6): pp. 505–15, 2009 Dec.

Brandtzaeg, P. Homeostatic impact of indigenous microbiota and secretory immunity. *Benef Microbes* **1**(3): pp. 211–27, 2010 Sept 1.

Brocks, J. J., Logan, G. A., Buick, R., Summons, R. E. Archean molecular fossils and the early rise of eukaryotes. *Science* **285**(5430): pp. 1033–6, 1999 Aug 13.

Bruzzese, E., Volpicelli, M., Squeglia, V., Bruzzese, D., Salvini, F., Bisceglia, M., Lionetti, P., Cinquetti, M., Iacono, G., Amarri, S., Guarino A. A formula containing galacto- and fructo-oligosaccharides prevents intestinal and extra-intestinal infections: an observational study. *Clin Nutr* **28**(2): pp. 156–61, 2009.

Buckley, M. R. Microbial Communities: From Life Apart to Life Together. Washington, DC: American Academy of Microbiology. Available at: http://academy.asm.org/images/stories/documents/microbialcommunities.pdf, 2003.

Cani, P. D., Delzenne, N. M. Interplay between obesity and associated metabolic disorders: new insights into the gut microbiota. *Curr Opin Pharmacol* **9**, pp. 737–43, 2009.

Cani, P. D., Amar, J., Iglesias, M. A., Poggi, M., Knauf, C., Bastelica, D., et al. Metabolic endotoxemia initiates obesity and insulin resistance. *Diabetes* **56**(7): pp. 1761–72, 2007.

Cani, P. D., Bibiloni, R., Knauf, C., Waget, A., Neyrinck, A. M., Delzenne, N. M., Burcelin, R. Changes in gut microbiota control metabolic endotoxemia-induced inflammation in high-fat diet-induced obesity and diabetes in mice. *Diabetes* **57**(6): pp. 1470–81, 2008 Jun.

Cani, P. D., Possemiers, S., Van de, W. T., Guiot, Y., Everard, A., Rottier, O., et al. Changes in gut microbiota control inflammation in obese mice through a mechanism involving GLP-2-driven improvement of gut permeability. *Gut* **58**: pp. 1091–103, 2009.

Carol, M., Borruel, N., Antolin, M., Llopis, M., Casellas, F., et al. Modulation of apoptosis in intestinal lymphocytes by a probiotic bacteria in Crohn's disease. *J Leukoc Biol* **79**: pp. 917–22, 2006.

Cherbut, C. Motor effects of short-chain fatty acids and lactate in the gastrointestinal tract. *Proc Nutr Soc* **62**: pp. 95–9, 2003.

Conly, J. M., Stein, K., Worobetz, L., Rutledge-Harding, S. The contribution of vitamin K2 (metaquinones) produced by the intestinal microflora to human nutritional requirements for vitamin K. *Am J Gastroenterol* **89**: pp. 915–23, 1994.

Coppa, G. V., Bruni, S., Morelli, L., Soldi, S., Gabrielli, O. The first prebiotics in humans: human milk oligosaccharides. *J Clin Gastroenterol* **38**(6 Suppl): pp. S80–3, 2004.

Costello, E. K., Lauber, C. L., Hamady, M., Fierer, N., Gordon, J. I., Knight, R. Bacterial community variation in human body habitats across space and time. *Science* **326**(5960): pp. 1694–7, 2009.

Cryan, J. F., O'Mahony, S. M. The microbiome–gut–brain axis: from bowel to behavior. *Neurogastroenterol Motil* **23**(3): pp. 187–92, 2011 Mar.

Cummings, J. H. Short chain fatty acids in the human colon. *Gut* **22**: pp. 763–79, 1981.

Cummings, J. H., Antoine, J. M., Azpiroz, F., Bourdet-Sicard, R., Brandtzaeg, P., Calder, P. C., Gibson, G. R., Guarner, F., Isolauri, E., Pannemans, D., Shortt, C., Tuijtelaars, S., Watzl, B. PASSCLAIM—Gut health and immunity. *Eur J Nutr* **43**(Suppl. 2): pp. II/118–73, 2004.

De Preter, V., Vanhoutte, T., Huys, G., Swings, J., De Vuyst, L., Rutgeerts, P., Verbeke, K. Effects of *Lactobacillus casei* Shirota, *Bifidobacterium breve*, and oligofructose-enriched inulin on colonic nitrogen-protein metabolism in healthy humans. *Am J Physiol Gastrointest Liver Physiol* **292**: pp. G358–68, 2007.

De Preter, V., Vanhoutte, T., Huys, G., Swings, J., Rutgeerts, P., Verbeke, K. Baseline microbiota activity and initial bifidobacteria counts influence responses to prebiotic dosing in healthy subjects. *Aliment Pharmacol Ther* **27**: pp. 504–13, 2008.

Derrien, M., Vaughan, E. E., Plugge, C. M., de Vos, W. M. Akkermansia muciniphila gen. nov., sp. nov., a human intestinal mucin-degrading bacterium. *Int J Syst Evol Microbiol* **54**: pp. 1469–76, 2004.

Deshpande, G., Rao, S., Patole, S., Bulsara, M. Updated meta-analysis of probiotics for preventing necrotizing enterocolitis in preterm neonates. *Pediatrics* **125**(5): pp. 921–30, 2010 May.

Dethlefsen, L., Eckburg, P. B., Bik, E. M., Relman, D. A. Assembly of the human intestinal microbiota. *Trends Ecol Evol* **21**(9): pp. 517–23, 2006 Sept.

Dethlefsen, L., McFall-Ngai, M., Relman, D. A. An ecological and evolutionary perspective on human-microbe mutualism and disease. *Nature* **449**(7164): pp. 811–8, 2007 Oct 18.

Dethlefsen, L., Huse, S., Sogin, M. L., Relman, D. A. The pervasive effects of an antibiotic on the human gut microbiota, as revealed by deep 16S rRNA sequencing. *PLoS Biol* **6**(11): p. e280, 2008.

DIAMOND Project Group. Incidence and trends of childhood Type 1 diabetes worldwide 1990–1999. *Diabet Med* **23**(8): pp. 857–66, 2006 Aug.

Diaz-Heijtz, R., Wang, S., Anuar, F., Qian, Y., Björkholm, B., Samuelsson, A., Hibberd, M. L., Forssberg, H., Pettersson, S. Normal gut microbiota modulates brain development and behavior. *Proc Natl Acad Sci U S A* **108**(7): pp. 3047–52, 2011 Feb 15.

Domínguez-Bello, M. G., Costtello, E. K., Contreras, M., et al. Delivery mode shapes the acquisition and structure of the initial microbiota across multiple body habitats in newborns. *Proc Natl Acad Sci U S A* **107**: pp. 11971–5, 2010.

Dominguez-Bello, M. G., Blaser, M. J., Ley, R. E., Knight, R. Development of the human gastrointestinal microbiota and insights from high-throughput sequencing. *Gastroenterology* **140**(6): pp. 1713–9, 2011 May.

Eckburg, P. B., Bik, E. M., Bernstein, C. N., Purdom, E., Dethlefsen, L., Sargent, M., Gill, S. R., Nelson, K. E., Relman, D. A. Diversity of the human intestinal microbial flora. *Science* **308**: pp. 1635–8, 2005.

Egert, M., de Graaf, A. A., Smidt, H., de Vos, W. M., Venema, K. Beyond diversity: functional microbiomics of the human colon. *Trends Microbiol* **14**: pp. 86–91, 2006.

Erridge, C., Attina, T., Spickett, C. M., Webb, D. J. A high-fat meal induces low-grade endotoxemia: evidence of a novel mechanism of postprandial inflammation. *Am J Clin Nutr* **86**(5): pp. 1286–92, 2007.

Falk, P. G., Hooper, L. V., Midtvedt, T., Gordon, J. I. Creating and maintaining the gastrointestinal ecosystem: what we know and need to know from gnotobiology. *Microbiol Mol Biol Rev* **62**: pp. 1157–70, 1998.

Flint, H. J., Bayer, E. A., Rincon, M. T., Lamed, R., White, B. A. Polysaccharide utilization by gut bacteria: potential for new insights from genomic analysis. *Nat Rev Microbiol* **6**: pp. 121–31, 2008.

Foligne, B., Zoumpopoulou, G., Dewulf, J., Ben Younes, A., Chareyre, F., et al. A key role of dendritic cells in probiotic functionality. *PLoS ONE* **2**: p. e313, 2007.

Frank, D. N., St Amand, A. L., Feldman, R. A., et al. Molecular–phylogenetic characterization of microbial community imbalances in human inflammatory bowel diseases. *Proc Natl Acad Sci U S A* **104**: pp. 13780–5, 2007.

Frank, D. N., Pace, N. R. Gastrointestinal microbiology enters the metagenomics era. *Curr Opin Gastroenterol* **24**: pp. 4–10, 2008.

Gaboriau-Routhiau, V., Rakotobe, S., Lecuyer, E., Mulder, I., Lan, A., Bridonneau, C., et al. The key role of segmented filamentous bacteria in the coordinated maturation of gut helper T cell responses. *Immunity* **31**: pp. 677–89, 2009.

Gatt, M., Reddy, B. S., MacFie, J. Review article: bacterial translocation in the critically ill—evidence and methods of prevention. *Aliment Pharmacol Ther* **25**(7): pp. 741–57, 2007 Apr 1.

Gibson, G. R., Roberfroid, M. B. Dietary modulation of the human colonic microbiota: introducing the concept of prebiotics. *J Nutr* **125**: pp. 1401–12, 1995.

Gibson, G. R., Beatty, E. R., Wang, X., Cummings, J. H. Selective stimulation of Bifidobacteria in the human colon by oligofructose and inulin. *Gastroenterology* **108**: pp. 975–82, 1995.

Gibson, G. R., Probert, H. M., Van Loo, J., Rastall, R. A., Roberfroid, M. B. Dietary modulation of the human colonic microbiota: updating the concept of prebiotics. *Nutr Research Reviews* **17**: pp. 259–75, 2004.

Gill, S. R., M. Pop, R. T. DeBoy, P. B. Eckburg, P. J. Turnbaugh, B. S. Samuel, J. I. Gordon, D. A. Relman, C. M. Fraser-Liggett, and K. E. Nelson. Metagenomic analysis of the human distal gut microbiome. *Science* **312**(5778): pp. 1355–9, 2006.

Gionchetti, P., Rizzello, F., Venturi, A., Brigidi, P., Matteuzzi, D., Bazzocchi, G., Poggioli, G., Miglioli, M., Campieri, M. Oral bacteriotherapy as maintenance treatment in patients with chronic pouchitis: a double-blind, placebo-controlled trial. *Gastroenterology* **119**: pp. 305–9, 2000.

Guarner, C., Soriano, G. Bacterial translocation and its consequences in patients with cirrhosis. *Eur J Gastroenterol Hepatol* **17**(1): pp. 27–31, 2005 Jan.

Guarner, F., Malagelada, J. R. Gut flora in health and disease. *Lancet* **361**: pp. 512–9, 2003.

Guarner, F., Bourdet-Sicard, R., Brandtzaeg, P., Gill, H. S., McGuirk, P., van Eden, W., Versalovic, J., Weinstock, J. V., Rook, G. A. Mechanisms of Disease: the hygiene hypothesis revisited. *Nat Clin Pract Gastroenterol Hepatol* **3**: pp. 275–84, 2006.

Guarner, F. Hygiene, microbial diversity and immune regulation. *Curr Opin Gastroenterol* **23**: pp. 667–72, 2007.

Guarner, F. What is the role of the enteric commensal flora in IBD? *Inflamm Bowel Dis* **14**(S2): pp. S83–4, 2008 Sep 24.

Haller, D., Bode, C., Hammes, W. P., Pfeifer, A. M., Schiffrin, E. J., Blum, S. Non-pathogenic bacteria elicit a differential cytokine response by intestinal epithelial cell/leucocyte co-cultures. *Gut* **47**: pp. 79–87, 2000.

Hart, A. L., Lammers, K., Brigidi, P., Vitali, B., Rizzello, F., Gionchetti, P., et al. Modulation of human dendritic cell phenotype and function by probiotic bacteria. *Gut* **53**(11): pp. 1602–9, 2004.

Hooper, L. V., Macpherson, A. J. Immune adaptations that maintain homeostasis with the intestinal microbiota. *Nat Rev Immunol* **10**(3): pp. 159–69, 2010 Mar.

Hooper, L. V., Midtvedt, T. and Gordon, J. I. How host-microbial interactions shape the nutrient environment of the mammalian intestine. *Annu Rev Nutr* **22**: pp. 283–307, 2002.

Hooper, L. V., Wong, M. H., Thelin, A., Hansson, L., Falk, P. G., Gordon, J. I. Molecular analysis of commensal host-microbial relationships in the intestine. *Science* **291**: pp. 881–4, 2001.

Hou, Y., Lin, S. Distinct gene number- genome size relationships for eukaryotes and non-eukaryotes: gene content estimation for dinoflagellate genomes. *PLoS ONE* **4** (9): p. e6978, 2009.

Joint FAO/WHO Expert Consultation. Health and nutritional properties of probiotics in food including powder milk with live lactic acid bacteria. Available at: http://www.fao.org/ag/agn/agns/micro_probiotics_en.asp, 2001.

Kagnoff, M. F. Microbial-epithelial cell crosstalk during inflammation: the host response. *Ann NY Acad Sci* **1072**: pp. 313–20, 2006.

Kalliomaki, M., Collado, M. C., Salminen, S., Isolauri, E. Early differences in fecal microbiota composition in children may predict overweight. *Am J Clin Nutr* **87**(3): pp. 534–8, 2008.

Karaki, S., Tazoe, H., Hayashi, H., et al. Expression of the short-chain fatty acid receptor, GPR43, in the human colon. *J Mol Histol* **39**, pp. 135–42, 2008.

Kelly, D., Campbell, J. I., King, T. P., Grant, G., Jansson, E. A., Coutts, A. G., Pettersson, S., Conway, S. Commensal anaerobic gut bacteria attenuate inflammation by regulating nuclear-cytoplasmic shuttling of PPAR-gamma and RelA. *Nat Immunol* **5**: pp. 104–12, 2004.

Kruis, W., Fric, P., Pokrotnieks, J., Lukas, M., Fixa, B., Kascak, M., Kamm, M. A., Weismueller, J., Beglinger, C., Stolte, M., Wolff, C., Schulze, J. Maintaining remission of ulcerative colitis with the probiotic *Escherichia coli* Nissle 1917 is as effective as with standard mesalazine. *Gut* **53**: pp. 1617–23, 2004.

Landy, J., Al-Hassi, H. O., McLaughlin, S. D., Walker, A. W., Ciclitira, P. J., Nicholls, R. J., Clark, S. K., Hart, A. L. Review article: faecal transplantation therapy for gastrointestinal disease. *Aliment Pharmacol Ther*. **34**(4): pp. 409–15, 2011 Aug.

Langlands, S. J., Hopkins, M. J., Coleman, N., Cummings, J. H. Prebiotic carbohydrates modify the mucosa-associated microflora of the human large bowel. *Gut* 53: pp. 1610–6, 2004.

Le Poul, E., Loison, C., Struyf, S., et al. Functional characterization of human receptors for short chain fatty acids and their role in polymorphonuclear cell activation. *J Biol Chem* **278**, pp. 25481–9, 2003.

Lee, Y. K., Menezes, J. S., Umesaki, Y., Mazmanian, S. K. Proinflammatory T-cell responses to gut microbiota promote experimental autoimmune encephalomyelitis. *Proc Natl Acad Sci U S A* **108** Suppl 1: pp. 4615–22, 2011 Mar 15.

Lepage, P., Seksik, P., Sutren, M., de la Cochetiere, M. F., Jian, R., Marteau, P., Dore, J. Biodiversity of the mucosa-associated microbiota is stable along the distal digestive tract in healthy individuals and patients with IBD. *Inflamm Bowel Dis* **11**: pp. 473–80, 2005.

Levitt, M. D., Gibson, G. R., Christl, S. Gas metabolism in the large intestine. In: Gibson, G. R., Macfarlane, G. T., editors. *Human Colonic Bacteria: Role in Nutrition, Physiology and Health.* CRC Press. Boca Raton, USA. pp. 113–54, 1995.

Lewis, S., Burmeister, S., Brazier, J. Effect of the prebiotic oligofructose on relapse of Clostridium difficile-associated diarrhoea: a randomized, controlled study. *Clin Gastroenterol Hepatol* **3**, pp. 442–8, 2005.

Ley, R. E., Peterson, D. A., Gordon, J. I. Ecological and evolutionary forces shaping microbial diversity in the human intestine. *Cell* **124**: pp. 837–48, 2006a.

Ley, R. E., Turnbaugh, P. J., Klein, S., Gordon, J. I. Microbial ecology: human gut microbes associated with obesity. *Nature* **444**: pp. 1022–23, 2006b.

Lichtman, S. M. Baterial translocation in humans. *J Ped Gastroenterol Nutr* **33**: pp. 1–10, 2001.

Llopis, M., Antolin, M., Carol, M., Borruel, N., Casellas, F., et al. Lactobacillus casei downregulates commensals' inflammatory signals in Crohn's disease mucosa. *Inflamm Bowel Dis* **15**: pp. 275–83, 2009.

Llopis, M., Antolin, M., Guarner, F., Salas, A., Malagelada, J. R. Mucosal colonization with Lactobacillus casei mitigates barrier injury induced by exposure to trinitronbenzene sulfonic acid. *Gut* **54**: pp. 955–9, 2005.

Lomax, A. R., Calder, P. C. Prebiotics, immune function, infection and inflammation: a review of the evidence. *Br J Nutr* **101**, pp. 633–58, 2009.

MacDonald, T. T., Monteleone, I., Fantini, M. C., Monteleone, G. Regulation of homeostasis and inflammation in the intestine. *Gastroenterology* **140**(6): pp. 1768–75, 2011 May.

Macfarlane, G. T., Gibson, G. R., Cummings, J. H. Comparison of fermentation reactions in different regions of the human colon. *J Appl Bacteriol* **72**: pp. 57–64, 1992.

MacFie, J. Current status of bacterial translocation as a cause of surgical sepsis. *Br Med Bull* **71**: pp. 1–11, 2004 Dec 13.

Macpherson, A., Khoo, U. Y., Forgacs, I., Philpott-Howard, J., Bjarnason, I. Mucosal antibodies in inflammatory bowel disease are directed against intestinal bacteria. *Gut* **38**: pp. 365–75, 1996.

Mai, V., Braden, C. R., Heckendorf, J., Pironis, B., Hirshon, J. M. Monitoring of stool microbiota in subjects with diarrhea indicates distortions in composition. *J Clin Microbiol* **44**(12): pp. 4550–2, 2006 Dec.

Mai, V., Young, C. M., Ukhanova, M., Wang, X., Sun, Y., Casella, G., Theriaque, D., Li, N., Sharma, R., Hudak, M., Neu, J. Fecal microbiota in premature infants prior to necrotizing enterocolitis. *PLoS One* **6**(6): p. e20647, 2011.

Manichanh, C., Rigottier-Gois, L., Bonnaud, E., Gloux, K., Pelletier, E., Frangeul, L., Nalin, R., Jarrin, C., Chardon, P., Marteau, P., Roca, J., Dore, J. Reduced diversity of faecal microbiota in Crohn's disease revealed by a metagenomic approach. *Gut* **55**: pp. 205–11, 2006.

Marteau, P., Rochart, P., Dore, J., Bera-Maillet, C., Bernalier, A., Corthier, G. Comparative study of bacterial groups within the human cecal and fecal microbiota. *Appl Environ Microbiol* **67**: pp. 4939–42, 2001.

Marteau, P., Lepage, P., Mangin, I., Suau, A., Dore, J., Pochart, P., Seksik, P. Review article: gut flora and inflammatory bowel disease. *Aliment Pharmacol Ther* **20** Suppl 4: pp. 18–23, 2004.

Martinez, C., Antolin, M., Santos, J., Torrejon, A., Casellas, F., Borruel, N., Guarner, F., Malagelada, J. R. Unstable composition of the fecal microbiota in ulcerative colitis during clinical remission. *Am J Gastroenterol* **103**: pp. 643–8, 2008.

Maslowski, K. M., Vieira, A. T., Ng, A., et al. Regulation of inflammatory responses by gut microbiota and chemoattractant receptor GPR43. *Nature* **461**, pp. 1282–6, 2009.

Mazmanian, S. K., Round, J. L., Kasper, D. L. A microbial symbiosis factor prevents intestinal inflammatory disease. *Nature* **453**(7195): pp. 620–5, 2008 May 29.

Metchnikoff, E. *The Prolongation of Life: Optimistic Studies.* Heinemann. London, United Kingdom, 1907.

Metges, C. C., Eberhard, M., Petzke, K. J. Synthesis and absorption of intestinal microbial lysine in humans and non-ruminant animals and impact on human estimated average requirement of dietary lysine. *Curr Opin Clin Nutr Metab Care* **9**(1): pp. 37–41, 2006 Jan.

Moayyedi, P., Ford, A. C., Talley, N. J., Cremonini, F., Foxx-Orenstein, A. E., Brandt, L. J., Quigley, E. M. The efficacy of probiotics in the treatment of irritable bowel syndrome: a systematic review. *Gut* **59**(3): pp. 325–32, 2010 Mar.

Moran, N. A. Symbiosis as an adaptive process and source of phenotypic complexity. *Proc Natl Acad Sci U S A* **104** Suppl 1: pp. 8627–33, 2007 May 15.

Moran, N. A., McCutcheon, J. P., Nakabachi, A. Genomics and evolution of heritable bacterial symbionts. *Annu Rev Genet* **42**: pp. 165–90, 2008.

Mullard, A. Microbiology: the inside story. *Nature* **453**(7195): 578–80, 2008 May 29.

Neufeld, K. M., Kang, N., Bienenstock, J., Foster, J. A. Reduced anxiety-like behavior and central neurochemical change in germ-free mice. *Neurogastroenterol Motil* **23**(3): pp. 255–64, e119, 2011 Mar.

O'Hara, A. M., Shanahan, F. The gut flora as a forgotten organ. *EMBO Rep* **7**: pp. 688–93, 2006.

O'Hara, A. M., Shanahan, F. Gut microbiota: mining for therapeutic potential. *Clin Gastroenterol Hepatol* **5**(3): pp. 274–84, 2007 Mar.

Pace, N. R. Mapping the tree of life: progress and prospects. *Microbiol Mol Biol Rev* **73**: pp. 565–76, 2009.

Palmer, C., Bik, E. M., Digiulio, D. B., Relman, D. A., Brown, P. O. Development of the human infant intestinal microbiota. *PLoS Biol* **5**: p. e177, 2007.

Pasteur, L. Observations relatives à la note précédente de M. Duclaux. *CR Acad Sci (Paris)* **100**: p. 68, 1885.

Pearce, N., Aït-Khaled, N., Beasley, R., Mallol, J., Keil, U., Mitchell, E., Robertson, C. and the ISAAC Phase Three Study Group. Worldwide trends in the prevalence of asthma symptoms: phase III of the International Study of Asthma and Allergies in Childhood (ISAAC). *Thorax* 2007 Sept; **62**(9): pp. 758–66. Epub 2007 May 15.

Pimentel, M., Lembo, A., Chey, W. D., Zakko, S., Ringel, Y., Yu, J., Mareya, S. M., Shaw, A. L., Bortey, E., Forbes, W. P., TARGET Study Group. Rifaximin therapy for patients with irritable bowel syndrome without constipation. *N Engl J Med* **364**(1): pp. 22–32, 2011 Jan 6.

Qin, J., Li, R., Raes, J., Arumugam, M., Burgdorf, K. S., Manichanh, C., Nielsen, T., Pons, N., Levenez, F., Yamada, T., Mende, D. R., Li, J., Xu, J., Li, S., Li, D., Cao, J., Wang, B., Liang, H., Zheng, H., Xie, Y., Tap, J., Lepage, P., Bertalan, M., Batto, J. M., Hansen, T., Le Paslier, D., Linneberg, A., Nielsen, H. B., Pelletier, E., Renault, P., Sicheritz-Ponten, T., Turner, K., Zhu, H., Yu, C., Li, S., Jian, M., Zhou, Y., Li, Y., Zhang, X., Li, S., Qin, N., Yang, H., Wang, J., Brunak, S., Doré, J., Guarner, F., Kristiansen, K., Pedersen, O., Parkhill, J., Weissenbach, J., MetaHIT Consortium, Bork, P., Ehrlich, S. D., Wang, J. A human gut microbial gene catalogue established by metagenomic sequencing. *Nature* **464**: pp. 59–65, 2010.

Rafter, J., Govers, M., Martel, P., Pannemans, D., Pool-Zobel, B., Rechkemmer, G., Rowland, I., Tuijtelaars, S., van Loo, J. PASSCLAIM—diet-related cancer. *Eur J Nutr* **43** Suppl 2: pp. 1147–84, 2004.

Reid, G., Sanders, M. E., Gaskins, H. R., Gibson, G. R., Mercenier, A., Rastall, R., Roberfroid, M., Rowland, I., Cherbut, C., Klaenhammer, T. R. New scientific paradigms for probiotics and prebiotics. *J Clin Gastroenterol* **37**: pp. 105–18, 2003.

Roberfroid, M. *Inulin-type Fructans: Functional Food Ingredients*. CRC Press. Boca Raton, USA, 2005.

Roberfroid, M., Gibson, G. R., Hoyles, L., McCartney, A. L., Rastall, R., Rowland, I., Wolvers, D., Watzl, B., Szajewska, H., Stahl, B., Guarner, F., Respondek, F., Whelan, K., Coxam, V., Davicco, M. J., Léotoing, L., Wittrant, Y., Delzenne, N. M., Cani, P. D., Neyrinck, A. M., Meheust, A. Prebiotic effects: metabolic and health benefits. *Br J Nutr* **104** Suppl 2: pp. S1–63, 2010.

Roller, M., Rechkemmer, G., Watzl, B. Prebiotic inulin enriched with oligofructose in combination with the probiotics *Lactobacillus rhamnosus* and *Bifidobacterium lactis* modulates intestinal immune functions in rats. *J Nutr* **134**(1): pp. 153–6, 2004.

Romagnani, S. Regulation of the development of type 2 T-helper cells in allergy. *Curr Opin Immunol* **6**: pp. 838–46, 1994.

Rook, G. A., Brunet, L. R. Microbes, immunoregulation, and the gut. *Gut* **54**(3): pp. 317–20, 2005.

Round, J. L., Mazmanian, S. K. The gut microbiota shapes intestinal immune responses during health and disease. *Nat Rev Immunol* **9**: pp. 313–23, 2009.

Round, J. L., Mazmanian, S. K. Inducible Foxp3+ regulatory T-cell development by a commensal bacterium of the intestinal microbiota. *Proc Natl Acad Sci U S A* **107**: pp. 12204–9, 2010.

Rowland, I. R. The role of the gastrointestinal microbiota in colorectal cancer. *Curr Pharm Des* **15**(13): pp. 1524–7, 2009.

Sanderson, I. R. Dietary modulation of GALT. *J Nutr* **137**, pp. 2557S–62S, 2007.

Sartor, R. B. Microbial influences in inflammatory bowel diseases. *Gastroenterology* **134**(2): pp. 577–94, 2008.

Savage, D. C. Microbial ecology of the gastrointestinal tract. *Annu. Rev. Microbiol* **31**: pp. 107–33, 1977.

Savino, F., Cordisco, L., Tarasco, V., Palumeri, E., Calabrese, R., Oggero, R., Roos, S., Matteuzzi, D. *Lactobacillus reuteri* DSM 17938 in infantile colic: a randomized, double-blind, placebo-controlled trial. *Pediatrics* **126**: pp. e526–33, 2010.

Sazawal, S., Hiremath, G., Dhingra, U., Malik, P., Deb, S., Black, R. E. Efficacy of probiotics in prevention of acute diarrhoea: a meta-analysis of masked, randomised, placebo-controlled trials. *Lancet Infect Dis* **6**: pp. 374–82, 2006.

Scholtens, P. A., Alliet, P., Raes, M., Alles, M. S., Kroes, H., Boehm, G., et al. Fecal secretory immunoglobulin A is increased in healthy infants who receive a formula with short-chain galacto-oligosaccharides and long-chain fructo-oligosaccharides. *J Nutr* **138**: pp. 1141–7, 2008.

Schopf, J. W. Microfossils of the Early Archean Apex chert: new evidence of the antiquity of life. *Science* **260**, pp. 640–6, 1993.

Schopf, J. W. Fossil evidence of Archaean life. *Philos Trans R Soc Lond B Biol Sci* **361**: pp. 869–85, 2006.

Schwiertz, A., Gruhl, B., Lobnitz, M., Michel, P., Radke, M., Blaut, M. Development of the intestinal bacterial composition in hospitalized preterm infants in comparison with breast-fed, full-term infants. *Pediatr Res* **54**: pp. 393–9, 2003.

Seksik, P., Rigottier-Gois, L., Gramet, G., Sutren, M., Pochart, P., Marteau, P., Jian, R., Doré, J. Alterations of the dominant faecal bacterial groups in patients with Crohn's disease of the colon. *Gut* **52**: pp. 237–42, 2003.

Shapiro, J. A. Thinking about bacterial populations as multicellular organisms. *Annu. Rev. Microbiol* **52**: pp. 81–104, 1998.

Sherman, M. P. New concepts of microbial translocation in the neonatal intestine: mechanisms and prevention. *Clin Perinatol* **37**(3): pp. 565–79, 2010 Sept.

Simon, G. L., Gorbach, S. L. Intestinal flora in health and disease. *Gastroenterology* **86**: pp. 174–93, 1984.

Sokol, H., Pigneur, B., Watterlot, L., et al. *Faecalibacterium prausnitzii* is an anti-inflammatory commensal bacterium identified by gut microbiota analysis of Crohn disease patients. *Proc Natl Acad Sci U S A* **105**: pp. 16731–6, 2008.

Stewart, I., Falconer, I. R. Cyanobacteria and cyanobacterial toxins. 'Oceans and Human Health: Risks and Remedies from the Seas,' pp. 271–96, Eds: Walsh, P. J., Smith, S. L. and Fleming, L. E. Academic Press, 2008.

Strachan, D. P., Taylor, E. M., Carpenter, R. G. Family structure, neonatal infection, and hay fever in adolescence. *Arch Dis Child* **74**: pp. 422–6, 1996.

Strober, W., Fuss, I., Mannon, P. The fundamental basis of inflammatory bowel disease. *J Clin Invest* **117**: pp. 514–21, 2007.

Suau, A., Bonnet, R., Sutren, M., Godon, J. J., Gibson, G., Collins, M. D., Dore, J. Direct rDNA community analysis reveals a myriad of novel bacterial lineages within the human gut. *Appl Environ Microbiol* **65**: pp. 4799–807, 1999.

Swidsinski, A., Ladhoff, A., Pernthaler, A., Swidsinski, S., Loening-Baucke, V., Ortner, M., Weber, J., Hoffmann, U., Schreiber, S., Dietel, M., Lochs, H. Mucosal flora in inflammatory bowel disease. *Gastroenterology* **122**: pp. 44–54, 2002.

Szajewska, H., Skorka, A., Dylag, M. Meta-analysis: *Saccharomyces boulardii* for treating acute diarrhoea in children. *Aliment Pharmacol Ther* **25**: pp. 257–64, 2007.

Telford, G., Wheeler, D., Williams, P., Tomkins, P. T., Appleby, P., Sewell, H., Stewart, G. S., Bycroft, B. W., Pritchard, D. I. The *Pseudomonas aeruginosa* quorum-sensing signal molecule n-(3-oxododecanoyl)-l-homoserine lactone has immunomodulatory activity. *Infect Immun* **66**: pp. 36–42, 1998.

Thanbichler, M., Wang, S., Shapiro, L. The bacterial nucleoid: a highly organized and dynamic structure. *J Cell Biochem* **96** (3): pp. 506–21, 2005.

Tilg, H., Kaser, A. Gut microbiome, obesity, and metabolic dysfunction. *J Clin Invest* **121**(6): pp. 2126–32, 2011.

Tong, J. L., Ran, Z. H., Shen, J., Zhang, C. X., Xiao, S. D. Meta-analysis: the effect of supplementation with probiotics on eradication rates and adverse events during *Helicobacter pylori* eradication therapy. *Aliment Pharmacol Ther* **25**: pp. 155–68, 2007.

Turnbaugh, P. J., Ley, R. E., Hamady, M., Fraser-Liggett, C. M., Knight, R., Gordon, J. I. The human microbiome project. *Nature* **449**(7164): pp. 804–10, 2007 Oct 18.

Turnbaugh, P. J., Hamady, M., Yatsunenko, T., Cantarel, B. L., Duncan, A., et al. A core gut microbiome in obese and lean twins. *Nature* **457**: pp. 480–84, 2009.

Tursi, A., Brandimarte, G., Papa, A., Giglio, A., Elisei, W., Giorgetti, G. M., Forti, G., Morini, S., Hassan, C., Pistoia, M. A., Modeo, M. E., Rodino, S., D'Amico, T., Sebkova, L., Sacca, N., Di Giulio, E., Luzza, F., Imeneo, M., Larussa, T., Di Rosa, S., Annese, V., Danese, S., Gasbarrini, A. Treatment of relapsing mild-to-moderate ulcerative colitis with the probiotic VSL#3 as adjunctive to a standard pharmaceutical treatment: a double-blind, randomized, placebo-controlled study. *Am J Gastroenterol* **105**(10): pp. 2218–27, 2010 Oct.

Van Baarlen, P., Troost, F., van der Meer, C., Hooiveld, G., Boekschoten, M., Brummer, R. J., Kleerebezem, M. Human mucosal in vivo transcriptome responses to three lactobacilli indicate how probiotics may modulate human cellular pathways. *Proc Natl Acad Sci U S A* **108** Suppl 1: pp. 4562–9, 2011 Mar 15.

Van Loo, J., Coussement, P., de Leenheer, L., Hoebregs, H., Smits, G. On the presence of inulin and oligofructose as natural ingredients in the western diet. *Crit Rev Food Sci Nutr* **35**: pp. 525–52, 1995.

Van Nimwegen, F. A., Penders, J., Stobberingh, E. E., Postma, D. S., Koppelman, G. H., Kerkhof, M., Reijmerink, N. E., Dompeling, E., van den Brandt, P. A., Ferreira, I., Mommers, M., Thijs, C. Mode and place of delivery, gastrointestinal microbiota, and their influence on asthma and atopy. *J Allergy Clin Immunol* PMID: 21872915, 2011 Aug 26.

Waldram, A., Holmes, E., Wang, Y., Rantalainen, M., Wilson, I. D., Tuohy, K. M., et al. Top-down systems biology modeling of host metabotype-microbiome associations in obese rodents. *J Proteome Res* **8**(5): pp. 2361–75, 2009.

Waligora-Dupriet, A. J., Campeotto, F., Nicolis, I., et al. Effect of oligofructose supplementation on gut microflora and well-being in young children attending a day care centre. *Int J Food Microbiol* **113**, pp. 108–13, 2007.

Walker, R., Buckley, M. Probiotic microbes: the scientific basis. American Academy of Microbiology. Available at: http://academy.asm.org/index.php/colloquium-program/clinical-medical-a-public-health-microbiology-/173-probiotic-microbes-the-scientific-basis-june-2006a, 2006.

Walter, J., Ley, R. E. The human gut microbiome: Ecology and recent evolutionary changes. *Annu Rev Microbiol* PMID: 21682646, 2011 Jun 16

Williams, E. A., Coxhead, J. M., Mathers, J. C. Anti-cancer effects of butyrate: use of micro-array technology to investigate mechanisms. *Proc Nutr Soc* **62**: pp. 107–15, 2003.

Wostmann, B. S. The germfree animal in nutritional studies. *Annu Rev Nutr* **1**: pp. 257–79, 1981.

Wu, G. D., Chen, J., Hoffmann, C., Bittinger, K., Chen, Y. Y., Keilbaugh, S. A., Bewtra, M., Knights, D., Walters, W. A., Knight, R., Sinha, R., Gilroy, E., Gupta, K., Baldassano, R.,

Nessel, L., Li, H., Bushman, F. D., Lewis, J. D. Linking long-term dietary patterns with gut microbial enterotypes. *Science* [Epub ahead of print], 2011 Sep 1.

Yamanaka, T., Helgeland, L., Farstad, I. N., Fukushima, H., Midtvedt, T., Brandtzaeg, P. Microbial colonization drives lymphocyte accumulation and differentiation in the follicle-associated epithelium of Peyer's patches. *J Immunol* **170**: pp. 816–22, 2003.

Zoetendal, E. G., Rajilic-Stojanovic, M., de Vos, W. M. High-throughput diversity and functionality analysis of the gastrointestinal tract microbiota. *Gut* **57**(11): pp. 1605–15, 2008.

Zoetendal, E. G., Akkermans, A. D., Akkermans-van Vliet, W. M., de Visser, J. A., de Vos, W. M. The host genotype affects the bacterial community in the human gastrointestinal tract. *Microb Ecol Health Dis* **13**: pp. 129–34, 2001.

Zoetendal, E. G., vom Wright, A., Vilpponen-Salmela, T., Ben-Amor, K., Akkermans, A. D. L., de Vos, W. M. Mucosa-associated bacteria in the human gastrointestinal tract are uniformly distributed along the colon and differ from the community recovered from feces. *Appl Environ Microbiol* **68**: pp. 3401–7, 2002.

Author Biography

Dr. Francisco Guarner graduated in Medicine at the University of Barcelona in 1973. He trained in Internal Medicine, Gastroenterology and Hepatology at the Hospital Clinic in Barcelona with Prof. Joan Rodes and at the University Clinic of Navarra with Prof. Jesus Prieto. His Ph.D degree in Medicine was obtained at the University of Navarra in 1983 for his research studies on liver cell cytoprotection with prostaglandins. He has been Visiting Scientist and Research Fellow at the Upjohn Company in Kalamazoo (Michigan, 1983), the Royal Free Hospital (London, 1984), the King's College Hospital (London, 1984–1986), and the Wellcome Research Laboratories (Beckenham, 1986). He joined the Digestive System Research Unit in Hospital Vall d'Hebron in 1987.

Dr. Francisco Guarner is currently Consultant of Gastroenterology at the Digestive System Research Unit and Head of the Experimental Laboratory in University Hospital Vall d'Hebron (Barcelona). He is a member of the Scientific Committee of the Research Institution of University Hospital Vall d'Hebron (www.vhir.org), member of the Board of Directors of the International Scientific Association for Probiotics and Prebiotics (www.ISAPP.net), and member of the Steering Committee of the International Human Microbiome Consortium (IHMC, www.human-microbiome.org). He is co-author of 300 publications on original research or reviews in the field of mucosal immunity, gastrointestinal inflammation and gut microbiota.

